Taking Flight

Taking Flight

A History of Birds and People in the Heart of America

Michael Edmonds

Wisconsin Historical Society Press

Published by the Wisconsin Historical Society Press
Publishers since 1855

The Wisconsin Historical Society helps people connect to the past by collecting, preserving, and sharing stories. Founded in 1846, the Society is one of the nation's finest historical institutions.
Join the Wisconsin Historical Society: wisconsinhistory.org/membership

Printed in the United States of America
Designed by Ryan Scheife / Mayfly Design

22 21 20 19 18 1 2 3 4 5

Library of Congress Cataloging-in-Publication Data
Names: Edmonds, Michael, 1952– author.
Title: Taking flight : a history of birds and people in the heart of America / Michael Edmonds.
Description: Madison, WI : Wisconsin Historical Society Press, [2017] | Includes bibliographical references.
Identifiers: LCCN 2017024588| ISBN 9780870208362 (paperback : alk. paper) | ISBN 9780870208379 (ebook)
Subjects: LCSH: Birds—Middle West—History.
Classification: LCC QL672.73.U6 E36 2017 | DDC 598.0977—dc23 LC record available at https://lccn.loc.gov/2017024588

♾ The paper used in this publication meets the minimum requirements of the American National Standard for Information Sciences—Permanence of Paper for Printed Library Materials, ANSI Z39.48-1992.

We are what we think.
All that we are arises with our thoughts.
With our thoughts we create the world.

BUDDHA

It is theory which first determines what can be observed.

ALBERT EINSTEIN

Publication of this book was made possible in part through generous gifts from Tod Highsmith and Robert Dohmen.

THE CALL BY STEVE WAGNER

Contents

1

Thinking about Birds

I began birding more than thirty years ago, when my wife was pregnant for the first time. Anxious about my upcoming responsibilities, I did what many new fathers do. I ran for the hills—or, more accurately, for the swamps—where I distracted myself from impending fatherhood by trying to identify momentary bursts of color in Roger Tory Peterson's *Field Guide to the Birds*.

Mary and I had traveled that summer to a cabin poised on the edge of a northern lake. I was unpacking the car when a great blue heron floated silently past, almost close enough to touch. I'd never seen one before, never seen anything like it. When it landed on a small island just across from the cabin and stared back at me, I was hooked.

I later learned that almost every novice birder receives an invitation like that, and, as John Burroughs wrote in 1868, "The thrill of delight that accompanies it, and the feeling of fresh, eager inquiry that follows, can hardly be awakened by any other pursuit . . . There is a fascination about it quite overpowering."[1] I began spending as much time as I could outdoors with my eyes on the sky.

I came home before the baby arrived, of course, but not before I'd joined the ranks of nearly fifty million American bird-watchers. That's how many of us head outdoors armed with binoculars or peer at birds on backyard feeders each year. In addition, more than two million hunters brave freezing winds every fall to shoot waterfowl and game birds in the United States. And hundreds of professional ornithologists study birds in universities, museums, and zoos.[2]

Birds are all around us. They paddle along our shores, flee from our footsteps, perch on our fences, and soar above our houses. We see birds every time we step outside.

And we usually see what we expect to see. A wood duck perched on a low branch by a pond is a check mark for a birder, data to a biologist, a potential trophy for a hunter, and a source of wonder to a child. As an old Hindu proverb puts it, "When a pickpocket encounters a saint, all he notices are pockets." When we look at birds, we see what we look for. So did our ancestors.

Humans have inhabited the Midwest—the Great Lakes and the watersheds of the Ohio, upper Mississippi, and lower Missouri Rivers—for thousands of years. Our predecessors in America's heartland shared this landscape with nearly all of the same birds that we do but saw them through very different eyes, which prompted them to act in very different ways. This book describes the changing ways in which people have thought about and acted toward birds over the last twelve thousand years.

Many of our ancestors believed that if a whip-poor-will landed on your roof, death or bad luck was sure to follow. They thought swallows hiber-

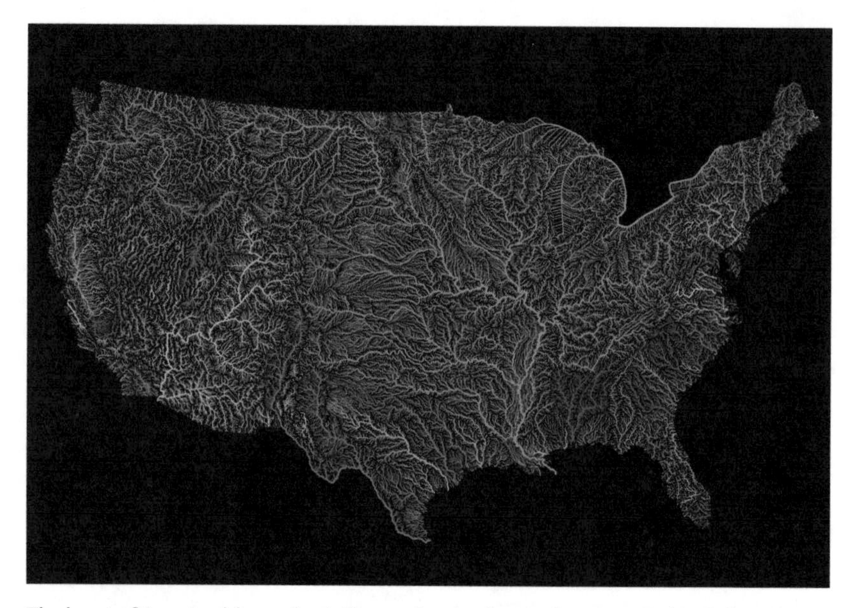

The heart of America (shown in pink) on a river basin map by cartographer Robert Szucs.
WWW.ETSY.COM/SHOP/GRASSHOPPERGEOGRAPHY

nated all winter in the mud beneath frozen ponds like frogs. They believed birds possessed souls, climbed high or dove deep into other realms, and could carry messages from the spirit world. They roasted vultures to smear the dripping fat on achy joints. They buried rooster claws under enemies' doorsteps to bring bad luck. They wore eagle feathers for courage and prayed to bird spirits for guidance. They held these views with the same certainty that we ascribe to scientific facts today. Like us, they saw their beliefs, desires, and values reflected in the birds that surrounded them.

For example, early white settlers regarded America's immense waves of passenger pigeons as free food and killed them recklessly. After all, their book of Genesis taught that God told them to "Rule over the fish in the sea and the birds in the sky and over every living creature that moves on the ground."[3] Three centuries later, the same biblical passage permitted rural hunters and urban diners to wipe out the last passenger pigeons.

But their American Indian neighbors saw passenger pigeons as children of the same Great Spirit that had created humans, as brethren, and honored them with ancient ceremonies before taking only as many as they needed. "Under our manner of securing them," recalled a nineteenth-century Seneca chief when the pigeons were nearly extinct, "they continued to increase."[4] People in both communities conformed to the unspoken expectations of their time and place.

Thirty years ago, when I distracted myself from impending fatherhood by taking up birding, I bought binoculars and a field guide. These placed me squarely within the tradition of Western science. The binoculars were a bequest from Galileo, the field guide a gift from Linnaeus. Together they reinforced my silent assumption that birds were just objects, specimens to be examined, labeled, and catalogued.

It took time for me to learn how to spot the diagnostic details that identify a bird, such as plumage, pattern, vocalizations, and bill shape, and even longer to master the bewildering taxonomy of Latin and English names. But by the end of that first summer, I could name most of the birds I saw in the field.

By mastering the language, I had joined a community with its own peculiar ideas and habits, a delightfully eccentric tribe that rises before dawn, stuffs thermoses into shoulder bags, wears ridiculous hats, and hikes through clouds of mosquitoes simply to look at birds. Joining local

Audubon Society outings and reading the classic books about birds, I embraced a shared way of seeing and thinking about nature.

As I scan the open water with my telescope on an icy December morning and spot a tundra swan near the opposite shore, I hold the bird at arm's length, so to speak. It is an object to be examined and identified. Tapping "tundra swan" into the iBird app on my cell phone, I instantly connect to three centuries of scientific observation and analysis. My reaction to the swan—what I perceive and how I process my perceptions—is a product of that history, which ignores some ways of knowing and privileges others. I can experience the swan only through a filter of specific beliefs, desires, and values.

This was driven home to me one morning when my five-year-old and I were playing with our dogs. When we stopped because it was time to get dressed for school, she asked, "Dad, do the dogs know what time it is?"

I stopped in my tracks. Of course the dogs didn't know what time it was. Time isn't something that can be sniffed or heard or seen. Without understanding numbers and the arithmetical patterns that govern them, the dogs could never know what time it was. I realized that I usually considered time to be part of nature outside myself, like air or gravity, when in fact it's actually a concept in my mind.

The dogs obviously couldn't know distance or direction, either. A foot, a meter, north, south, above, below, left, right—these, too, didn't exist outside me in the natural world. They were just ideas inside my head; every culture distinguishes them differently.

And, I realized, if such fundamental aspects of reality as time and space were socially constructed, then what about all the bird names I'd so carefully mastered? They weren't really about birds at all, but about the way that birders and scientists talk about birds.

"We cut up and organize the spread and flow of events as we do," wrote anthropologist Benjamin Lee Whorf, "largely because, through our mother tongue, we are parties to an agreement to do so, not because nature itself is segmented in exactly that way for all to see."[5] Language forms the banks within which our stream of consciousness flows.

Whorf urged us not to mistake scientific ideas for "the apex of the evolution of the human mind, nor their present wide spread as due to any survival from fitness, or to anything but a few events of history—events that could be called fortunate only from the parochial point of view of the favored parties."[6] American Indians around the Great Lakes speaking Ojibwe, eighteenth-century slaves in Missouri using remnants of West African languages, and the long-silenced peoples who lived here thousands of years before them all possessed their own unique ways of experiencing nature.

Each of us lives inside a web of shared assumptions, the things that everyone around us takes for granted. Some ideas seem like common sense, universally endorsed by public opinion, while others are literally unthinkable. Some actions seem right and proper, sanctioned by our laws and traditions, while we judge others to be so repulsive that only sociopaths perform them. This was also true for the Victorian market hunter in Illinois, the nineteenth-century voodoo queen in St. Louis, the eighteenth-century naturalist in Philadelphia, the seventeenth-century Jesuit in Wisconsin, the eleventh-century warrior in Missouri, and the third-century shaman in Ohio.

The world they left to us was created by the intersection of their ideas and their actions. Evidence of their actions is easy to see. It's all around us in nature, where passenger pigeons no longer darken the sky. We see it all across our landscape of dammed rivers, drained marshes, industrialized farms, and perpendicular highways.

However, our predecessors' ideas are harder to identify than their actions. The oldest were expressed in sacred objects that they buried with their dead, petroglyphs they carved into cave walls, and effigy mounds they sculpted on hilltops. In recent years, scholarly consensus has emerged about the beliefs expressed by the archaeological record, so it's possible to glimpse the many different things that birds meant to the peoples who preceded us here thousands of years ago. This is discussed in chapters two through four.

Ideas about birds are also preserved in centuries-old books and manuscripts in our great libraries. The first detailed descriptions of Midwestern birds were put on paper during the Renaissance by explorers and missionaries. They were followed by four centuries of nature writing by soldiers, travelers, and scientists. Their stories are told in chapters five through seven.

But the well-educated were not the only people fascinated by birds. Illiterate slaves, uneducated homesteaders, immigrant laborers, provincial hunters, and others who wrote no books possessed their own forms of knowledge. It's preserved in the colloquial names they coined for birds and in the folktales and proverbs they shared. By investigating these sources, we can see not only what educated, elite Midwesterners thought about birds, but also what the vast majority of ordinary folk thought about them. Their voices are heard in chapters eight and nine.

Some of our predecessors' ideas and actions make us scratch our heads in disbelief. What in the world were they thinking? We may find ourselves amazed at the superstitions, awe, fear, greed, and desire through which they saw the same birds that we see. We might be astonished by the bizarre things they did. We may judge them through our own cultural filters and pity them for having been born before science revealed the truth about the world. But this closes more doors than it opens.

One of the central principles of ethnobiologists is to "respect the integrity, morality, and spirituality of the culture, traditions, and relationships" of their native hosts. They try to work "only in the holistic context, respectful of norms and belief systems of the relevant communities."[7]

What if we applied this approach to the history of science and nature? What if, instead of searching the past for nuggets of fact and assessing their reliability, we approached our ancestors with the humility of an anthropologist working in the rain forest or on the arctic tundra? What might we discover if we try to empathize with these earlier people rather than judge them? By trying to appreciate how they understood the same birds that soar past our binoculars or perch on our feeders, we might learn not just about the past but also about ourselves.

The pages that follow cast no verdicts on the accuracy of historical observations or the validity of ancient explanations. My goals are simply to share the astonishing ways that people in the American heartland thought about birds, to explain how those thoughts prompted them to act, and to see what light their ideas and behavior may shed on our own.

Mind comes before matter. We see only what we're prepared to see, and then we impose our beliefs, desires, and values onto nature through our actions. So did our ancestors. The world we inherited from them was shaped in this way, as is the world that we will leave to our children.

2

Bones, Stones, Mounds, and Magic

When we passed the Ugly Coyote Saloon in tiny Denzer, Wisconsin, I thought we must have missed the turn.

Mary and I had left home early and were nearly an hour off the interstate, winding through mist-covered hills, past sleepy farms and scrubby woods. We'd hardly seen another car for twenty minutes, and as clouds wrapped around the trees and barns we felt like we were burrowing deeper and deeper into some northern outpost that time forgot. We were searching for Natural Bridge State Park, not to admire its amazing sandstone arch but to examine what lay beneath it.

Just as I opened my phone to check the GPS, the park's entrance sign leaped out from behind a bend. There in the parking lot, tucking their jeans into heavy socks, were Wisconsin state archaeologist John Broihahn and his wife, Gabriele Lubach. It was eighty degrees and humid, so I'd decided against long pants. John looked at me disdainfully and said, "I wore shorts into the field. Once."

John is charged with monitoring the health of the state's protected archaeological sites, and he'd agreed to let me tag along while he visited Raddatz Rockshelter, in Sauk County. He called it "Warren Wittry's Time Machine," after the archaeologist who excavated it sixty years ago.

As Wittry's 1950s crew dug down, they seemed to go back in time, unearthing American Indian artifacts stacked ten feet deep in subterranean layers. Radiocarbon dating had just been invented, so Wittry was able to carefully map the history of Wisconsin back to 10,000 BCE. He found not only spearpoints and stone tools, but also bird bones that proved to be the oldest evidence of contact between people and birds in the Midwest.

Long before Socrates drank the hemlock or Egyptians built the pyramids, American Indians were catching and eating birds here.

Most of the birds we know today finished evolving from dinosaurs sixty-five million years ago. Humans took longer to arrive—Homo sapiens is just two hundred thousand years old. The earliest conclusive proof of people and birds coming together dates from Africa about 40,000 BCE, roughly the same time that the first language and art emerged.[8]

At that time, much of North America was covered in a glacier thousands of feet high. Until roughly 16,000 BCE, ice sheets blanketed nearly everything from the Dakotas to the Ohio River and northeast to New England. As the earth gradually warmed, the glacier retreated northward and today's rivers, lakes, hills, and valleys were scraped into the landscape.

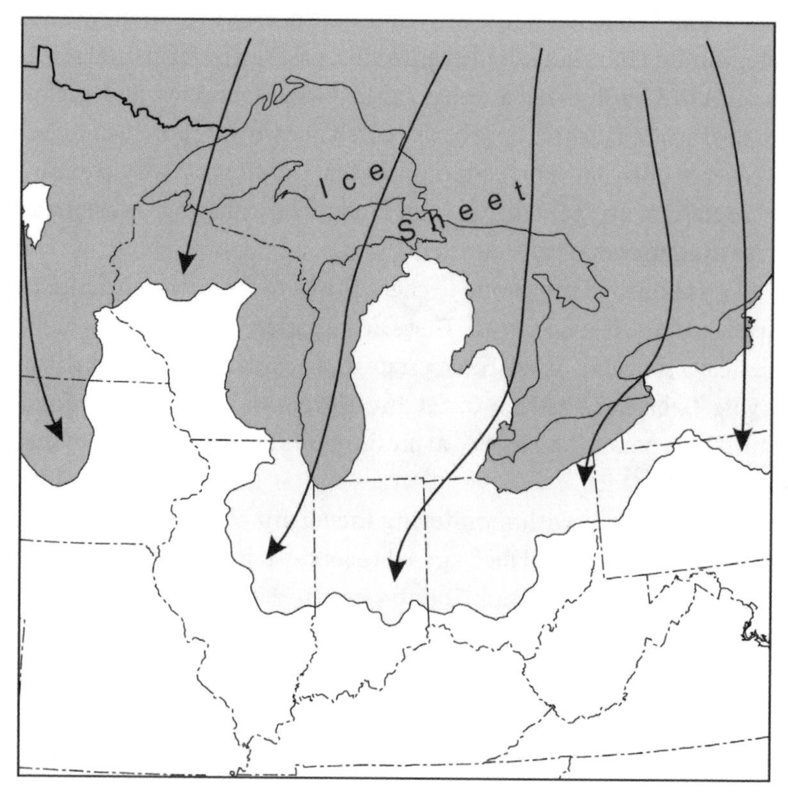

Extent of glaciation in the Midwest. HANSEL (2004), END MORAINES—THE END OF THE GLACIAL RIDE. ©2004 UNIVERSITY OF ILLINOIS BOARD OF TRUSTEES. USED WITH PERMISSION OF THE ILLINOIS STATE GEOLOGICAL SURVEY.

Most scientists believe that around 10,000 BCE, people moved from Alaska along the glacier's southern edge into newly exposed habitats that were cold, wet, and inhospitable. At selected spots that the ice sheets didn't reach in the Upper Mississippi and Ohio Valleys lie ancient caves like the Raddatz Rockshelter.[9]

As John, Gabriele, Mary, and I climbed the trail through prickly ash and underbrush, with me keeping an eye out for poison ivy and ticks on my bare legs, John told stories about the last two generations of Wisconsin archaeologists. How they dug and sifted earth in the burning sun all day and slept in canvas army surplus tents at night, laid out their survey grids with folding wooden rules instead of laser pointers, took notes with no. 2 pencils instead of iPads, and climbed onto rooftops at the solstice to experience the night sky as ancient Indians had seen it.

Cresting the hill, we stepped out of the woods into an open bowl about thirty yards wide. On our right, the towering sandstone bluff culminated in a natural bridge about twenty feet long and five feet wide spanning a gap the size of a dump truck. Beneath it, directly in front of us, lay a sloping cave roughly seven feet tall and thirty feet deep. In these shadows, American Indians had taken refuge from brutal Wisconsin winters more than twelve thousand years ago.

The people who lived here back then didn't grow food, wear cloth, make pottery, use metal, or even shoot bows and arrows. They moved from place to place in family bands of a few dozen people, killing large mammals with stone-tipped spears and collecting roots, fruits, nuts, and berries. During the summer, bands would meet together to arrange marriages, bury their dead, and perform ceremonies. Each fall, they dispersed again, hunkering down through Ice Age winters in protected spots like this one. They lived almost entirely off the flesh and hides of large mammals. Eighty-five percent of the 5,200 animal bones that Wittry found beneath Raddatz Rockshelter were from white-tailed deer. But he also found the bones of twenty-one species of birds.[10]

Most of the bird bones at the lowest level of Wittry's dig, dating to twelve thousand years ago, were from highly nutritious upland species like passenger pigeon and ruffed grouse. The first people at Raddatz clearly ate birds, but they were a statistically insignificant part of their mammal-based diet. Just a few bones of ducks and geese were found, al-

Raddatz Rockshelter, Natural Bridge State Park, Sauk County, Wisconsin, 2010. OFFICE OF THE STATE ARCHAEOLOGIST OF WISCONSIN

though back then a creek flooded low meadows at the bottom of the hill. This led archaeologists to conclude that Raddatz was probably occupied in the winter, after migrating waterfowl had departed in the fall and before they returned in the spring.

Besides grouse and pigeons, Wittry's crew found bones from woodpeckers, blue jays, and other small songbirds with bright plumage but minimal value as food. They also uncovered the large wing bones of hawks and other raptors, but no bones from other parts of their bodies. Both finds suggest that the people here used colorful feathers of small birds and hawk or eagle wings for religious or decorative purposes. They might also have exchanged them with other bands. A site in Illinois dating to six thousand years ago contained a large cache of isolated wing bones from geese and sandhill cranes. The wings, suitable for use as fans or ornaments, had traveled far from the rest of the birds' carcasses. The presence of such trading stockpiles suggests that people were doing more with birds than eating them from the earliest times of human occupation in the Midwest.[11]

Crew at work on Warren Wittry's excavation, Raddatz Rockshelter, Sauk County, Wisconsin, 1950s. WHS MUSEUM ARCHAEOLOGY PROGRAM

At Raddatz, bird remains were spread through all fifteen levels of the excavation, showing that people interacted with birds on the site throughout the course of the past twelve thousand years. In the early nineteenth century, Ho-Chunk residents explained to early white settlers that a second neighboring stone arch was sacred, and that the Ho-Chunk "never visited the camp without going to the natural bridge for worship, as that was regarded by them as 'a work of the Great Spirit.'" If ancient Indians felt the same way, they may have used colorful feathers in religious ceremonies here. Even white pioneers standing in the bowl beside the Raddatz Rockshelter found themselves moved. One early visitor doubted "whether a superior can be found in this whole region of country as a retired and romantic spot," and another called it "extremely fantastic."[12]

I had come to Raddatz hoping for a similar reaction, some intuition or half-conscious sense, at the very least, of how the first people to live in the heart of America understood the birds that surrounded them. I didn't find it. Generations of local teenagers have used this sacred place for drinking parties and carved their initials into its soft sandstone walls. Tourists hike up to look at the natural bridge, but few understand the significance of the rockshelter beneath it. The surrounding prehistoric meadows and wetlands have become scrubby woods and tidy cornfields. Raddatz contained no time machine for me.

As thunder rolled gently across the Baraboo Hills, we headed back to the parking lot. I had stood in the exact place where twelve thousand years ago some of America's first people had interacted with birds and had come away unmoved. I wondered if touching the actual bones unearthed there might bring me closer to the first people who engaged with Midwestern birds.

So on another sweltering August day, I descended into the cool basement of the Wisconsin Historical Society to examine the Raddatz bones with Janet Speth, one of a handful of avian archaeologists working in the United States. Two acid-free bankers boxes were retrieved from an underground storage vault for us and laid on a lab table. Inside each large box was a series of smaller ones, stored like Russian nesting dolls. Inside each small box were tiny bone-filled Ziploc bags, each labeled with a species name and the museum's accession number. These were the specimens dug from the earth and recovered in mesh filters by Warren Wittry in the 1950s.

"*Meleagris gallopavo*" read the label on the first bag. With a gentle twist of the wrist, Janet spread out half a dozen prehistoric turkey bones. They looked to me like typical Thanksgiving dinner debris, except for the minuscule names neatly written on them in black ink. As she picked up each dry bone, one after another, polysyllabic technical terms tumbled effortlessly off her tongue—metacarpal, distal end, mandible, humerus, and the prodigious tarsometatarsus. For an hour we went slowly through bag after bag, Janet trying unsuccessfully to initiate me into the mysteries of accelerator mass spectrometry, osteological indicators of climate change, and precisely how she could tell whether or not a bone had passed through the digestive tract of a dog.

She would have talked happily for a second hour, but her knowledge was wasted on me. I'm simply too ignorant to appreciate the depth of her expertise. Whenever I tried to rephrase it in ways that my layperson's mind could grasp, she would smile at me kindly, compassion and pity dancing together behind her lively gray eyes. The ancient bird bones, washed clean, neatly labeled, and wrapped in plastic, gave me no more than the rockshelter had. They didn't tell me anything meaningful about the people who had hunted and eaten these birds.

Perhaps I should have expected that. Philosopher Alan Watts once recalled his young children asking him to describe God. "I replied," he wrote, "that God is the deepest inside of everything." They were eating lunch at the time, and the children asked him to cut open a grape so they could see God. He obliged, but cutting the grape in half created two more outsides. No matter how many times he sliced a grape, he created more outsides. Whether crouching under Raddatz Rockshelter or handling the bones unearthed there, I, too, could see only outsides.[13]

Getting inside the heads of ancient people began to seem possible when I encountered artifacts made ten thousand years later (about 1500 BCE) when American Indians in the Midwest started to fashion birds into symbolic shapes. This roughly coincided with humans planting the first garden crops, inventing pottery, working with metal, living in semipermanent villages, and burying leaders in carefully constructed graves.

Trying to understand the beliefs of preliterate people by looking at the mute objects they left behind is risky. Most archaeologists prefer to stick closely to physical evidence, like the measurements of projectile points or pottery shards, and hesitate to make deductions about intellectual life from stones and bones.

This led archaeologist Robert Hall to worry in the 1970s that his profession was "more concerned about how Indians made their livings, than about what Indians thought it was worthwhile to live for." He and a few others in the field went beyond the physical evidence to also consider ethnological studies of modern American Indians, historical accounts from the recent past, insights from contemporary psychology, and even disciplines like linguistics and neurobiology. Equipped with these additional insights, they speculated about the beliefs of prehistoric peoples.[14]

Sometimes this interdisciplinary approach yields startling results. For example, anthropologist Peter Nabokov once told me about listening to archaeologists puzzling over a series of symbols repeated on a prehistoric pot, which he immediately recognized as steps in a dance that's still performed. Hall and other "cognitive archaeologists" argued that persistent motifs displayed on objects fabricated at different times express the same concepts in different media across the centuries, something like Jungian archetypes or Chomsky's universal grammar expressed in pots, tools, and rock art.

We can never test whether their conjectures are strictly true, of course, but it's more useful to speculate on what ancient people may have thought than to say nothing at all about their beliefs, desires, and values. Over the last thirty years, in fact, so many ancient decorative objects have been recovered and analyzed that a scholarly consensus has emerged about the likely worldview of our prehistoric predecessors in the Midwest.[15]

The first culture to leave abundant examples of artistic and intellectual life was the Adena, which flourished in the Ohio Valley circa 1200–100 BCE. Terms like "Adena culture" don't refer to a specific ethnic group with its own unique gene pool, language, and culture. Rather, archaeologists use them to categorize evidence of common behaviors that emerged, blossomed, and waned over certain periods in specific places. They connote something like a family resemblance, features found in peoples from different places who shared some essential habits but about whom we know almost nothing else.

Adena remains are clustered in the Ohio River Valley, but are also found in excavated sites stretching from western Pennsylvania to southern Indiana. From these centers, Adena residents traded with communities across much of the eastern United States, acquiring copper from as far away as Lake Superior and shells from the Atlantic Coast. Unlike the roaming Paleo-Indians who wintered at Raddatz ten thousand years earlier, they lived in semipermanent villages of round houses enclosed within sturdy earthwork walls.

Their garbage pits show that Adena peoples ate almost anything they could get their hands on—deer, bear, small mammals, turtles, snakes, freshwater mussels, fish, turkeys, trumpeter swans, songbirds—and also cultivated gardens with squash, pumpkins, and sunflowers. They wove plant fibers into fabric, made copper tools and utensils, and created intricate necklaces, bracelets, and headdresses. They placed the corpses of their leaders in wood or bark coffins alongside exquisite carvings and jewelry, and then covered them with conical mounds rising, in some cases, several stories tall.[16]

Anthropologists concluded from artifacts buried in Adena mounds that these people practiced shamanism and that birds were central to their religious life. Shamanism, in the words of archaeologist William Romain, is "a complex of beliefs and practices based on the notion that spirits pervade the cosmos and that these spirits can be personally contacted for specific purposes, through altered states of consciousness." It is found among indigenous people literally around the globe, from the Siberian Arctic to the Peruvian Amazon. Archaeological evidence suggests that it has been practiced since the dawn of humanity, and some anthropologists claim that shamanism is the source of all major world religions.[17]

For shamanists, what we can know through our five senses is just one of many realities. As twentieth-century Mexican shaman Maria Sabina put it, "There is a world beyond ours, a world that is far away, nearby, and invisible. And there is where God lives, where the dead live, the spirits and the saints, a world where everything has already happened, and everything is known. That world talks. It has a language of its own. I report what it says." Shamans in Ecuador believe that our normal waking consciousness "is simply a lie, or illusion, while the true forces that determine daily events are supernatural and can only be seen and manipulated with the aid of hallucinogenic drugs."[18]

Indigenous people everywhere have viewed the world this way for countless millennia. To shamanists (at the risk of oversimplification), the universe consists of an Upper World, part of it revealed to humans as the sky and stars; a Middle World that humans experience with their physical senses; and a Lower World, which can be entered through caves or under water. Passing through all three realms is an *axis mundi*, a cosmic spine on which shamans can travel when entranced. It is often depicted in ancient art as a Tree of Life with birds perched on its branches.

Souls of the dead and other spirit beings move across all three planes of existence, sometimes serving as personal guardians for humans or intervening in the material world in other ways. In this three-tiered universe, time unfolds cyclically, like Earth's seasons. Everything comes into existence, grows and develops, decays and dies, and then is reborn and renewed. The cosmos is timeless and everything possible has already happened. Everything humans encounter is sacred because the entire universe is animated by spiritual forces. Shamans cultivate specific techniques, including rituals involving birds, for moving through the two other realms in order to create outcomes in ours.[19]

"Whether in Asia, Australia, Africa, or North and South America," writes anthropologist Peter Furst,

> the shaman functions fundamentally in much the same way and with similar techniques—as guardian of the psychic and ecological equilibrium of his group and its members, as intermediary between the seen and unseen world, as master of spirits, as supernatural curer, etc. . . . He ensures, by means of his special powers and his unique psychological capacity to transcend the human condition and pass freely back-and-forth through the different cosmological planes (as these are conceived in the particular worldview of his group), the renewal of game animals and, indeed, of all nature.[20]

Shamans enter the other realms through trance, prayer, rituals like the sweat lodge, meditating on objects with sacred power, fasting, ecstatic dancing, sensory deprivation, repetitive chanting, and hallucinogenic plants. Neurobiologist Rick Strassman discovered in the 1990s that practices such as these can cause the brain's pineal gland to produce N,N-Dimethyltryptamine (DMT). When released in the brain, DMT produces visual and auditory hallucinations, out-of-body experiences, encounters with disembodied beings, and other altered states of consciousness. When ethnobotanist Terence McKenna gave synthetic DMT to shamans in the Amazon, they described the beings they met in their vision as "the spirits that we work with. These are ancestor souls. We know this place."[21]

Shamans in the heart of North America used several hallucinogenic plants in their ceremonies. The red bean or mescal bean (*Sophophora secundiflora*) was employed by the Potawatomi, Comanche, Iowa, Kansa, Omaha, Osage, Pawnee, Ponca, Otoe, and Wichita tribes during historic times to induce visions; archaeologists found it interred with ritual objects and depicted on rock art in a west Texas site dating to 8000 BCE. A twentieth-century Ojibwe shaman in the Great Lakes named Keewaydinoquay described how *Amanita muscaria*, or the fly agaric mushroom, has been used by her people for many generations to produce visions. Other hallucinogenic plants that grow in the Midwest or on the Great Plains include sweet flag (*Acorus calamus*), which, when chewed in large quantities, produces "visual hallucinations and other effects similar to those of LSD," jimsonweed (*Datura stramonium*), and morning glory (*Ipomea violacea*), the seeds of which produce psychedelic effects. Native tobacco species, which contain much higher concentrations of nicotine than our contemporary retail variety, were purposely cultivated for religious purposes.[22]

Reports from around the world over the last five centuries reveal that shamans wear masks, skins, skulls, and headdresses of animals whose powers they seek or with whom they hope to communicate. Through ritual purification, hypnotic drumming and chanting, ecstatic dancing, and hallucinogenic plants, shamans replace everyday consciousness with mystical states. Their followers believe that they actually take the physical form of other creatures or allow animal spirits to occupy their bodies. Birds are among their most important guides and helpers.[23]

"Birds," writes archaeologist Karine Tache of prehistoric people in the Midwest, "are powerful symbols perceived as mediators between the various cosmic realms. Bird motifs or anatomical parts—feathers, beaks, claws—are often part of a shaman's regalia and/or paraphernalia, conferring on him the power to communicate with creatures and spirits inhabiting realms beyond the world known to human beings." To these prehistoric people, birds also possessed transcendental powers that even laypeople could see: raptors soared so high that they disappeared into the Upper World, while loons and ducks vanished underwater to travel to the Lower one. To someone who sees the entire universe as a sacred whole, birds are obvious messengers between people and the gods.[24]

Twenty-five hundred years ago, Adena shamans in the Ohio Valley often employed birds in religious ceremonies. Their burial mounds contain hundreds of bird remains, images, and artifacts. The most remarkable are carved stone tablets dating from the first several hundred years BCE. Thirteen of these tablets have been unearthed, nine of which are incised with stylistic representations of birds.[25]

The so-called Berlin Tablet contains all the avian iconographic elements that appear sporadically on many other Adena artifacts. At the

Berlin Tablet with incised raptor from an Adena mound. COURTESY OF THE OHIO HISTORY CONNECTION, IMAGE A 340/000001

Wilmington Tablet. COURTESY OF THE OHIO HISTORY CONNECTION, A 3490/000210

upper right is a bird's head, with a long beak suggestive of a raptor. Below this, spanning the width of the oblong tablet, is the bird's body wrapped in an *S*-curve. At the lower right are its feet with prominent claws, and at the lower left, the wings. Finally, at the upper left is the bird's tail. In this beautifully symmetrical image, the bird's ferocity and power are emphasized even though it sits peacefully at rest.

Other Adena tablets, like the Wilmington Tablet, show pairs of birds as mirror images of each other. Anthropologists think this juxtaposition expressed the essential unity of opposites, like the Chinese yin-yang symbol, and reflected the cosmic wholeness where "everything has already happened, and everything is known."

Archaeologists suspect that a third plate, called the Lakin A Tablet, employs the same avian iconographic elements to show a shaman dancing in a bird mask. A human head is detached and placed to one side, as if to show that the shaman has left his or her body, and the arms and legs are feathered to convey that he or she has become a bird. The species most commonly shown on the Adena tablets are thought to be turkey vultures, which soar on thermals until they disappear from sight into the Upper World, and peregrine falcons, whose speed and ferocity are unmatched in the animal kingdom.[26]

Because the Adena tablets have been found only in shamans' graves, archaeologists know they were important ceremonial objects. And because some contain remnants of colored pigments in their grooves, researchers suspect the tablets were used to transfer bird designs to other surfaces. Pigment would have been daubed into the tablet's incised lines, its surface wiped clean, and the color then transferred to the recipient medium by pressing the two together. If true, these tablets are the oldest evidence of printing in North America. Although the Adena wove fabric from plant fibers, it was probably too coarse to receive these intaglio impressions. It's more likely that the designs on the tablets were printed onto soft, light-colored deerskin that would have easily absorbed them, or transferred as temporary tattoos directly onto human skin as part of a religious ritual or mortuary preparation.[27]

In addition to the incised tablets, archaeologists found beautiful stone sculptures of birds in Adena burial mounds. Most of these were probably not made in the Ohio Valley but traded from the southern shores of the

Lakin A Tablet. COURTESY OF THE OHIO HISTORY CONNECTION, A 4786/000081

eastern Great Lakes. They are found all across the Midwest, and archaeologists call them "birdstones."[28]

Birdstones are usually three to six inches long and were made in several shapes. The earliest, carved around 1000 BCE, are sleek, elongated forms with bowed heads and fan-shaped tails. A few hundred years later, variations were created with narrower tails and topknots on the heads. By 200 BCE, the bodies had become tall, blockish, and rectangular. Nearly all birdstones depict birds in resting poses and have two holes drilled through the base. Most were carved from banded slate, though a dozen other types of soft stone were also used. They were sanded perfectly smooth, probably with increasingly fine layers of grit.

Birdstones appear to have originated along the shores of the Great Lakes, in eastern Wisconsin, northern Illinois and Indiana, Ohio, Michigan, and western New York and Ontario. A few have been found as far west as Minnesota, as far east as New England, and as far south as Maryland. More than four thousand have been found in all. In Monroe County, Michigan, at the west end of Lake Erie, archaeologists unearthed an entire stone workshop with slate artifacts in varying stages of completion including birdstones, hammers, smoking pipes, gorgets (decorative plates worn on the chest), and other stone tools and ornaments.[29]

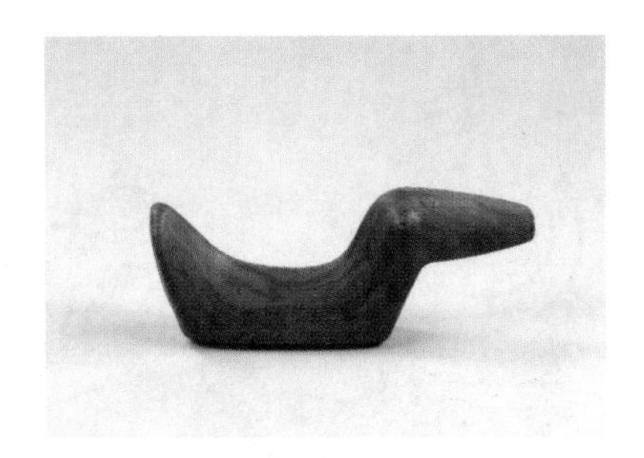

Birdstone from Sauk County, Wisconsin. HENRY P. HAMILTON COLLECTION, WHS MUSEUM OBJECT #1919.1385

Soon after the first birdstones were unearthed in the nineteenth century, their beauty made them collectors' items. In the decades that followed, the stones became so prized that counterfeits were sold in large numbers. Yet despite the popular interest, no one could figure out their original function.

Some scholars initially speculated they were ornaments worn on the head because some had been found in graves next to human skulls. Others imagined that they were fastened to flutes, to slide up and down over the holes to create notes. But the likeliest explanation is that they were used in hunting. In South and Central America, where spear-throwers were employed until historical times, birdstones and other effigies were often found attached to their handles.[30]

Put aside Hollywood images of natives in loincloths hurling six-foot spears by hand. Indigenous peoples in the Americas used a tool called an atlatl that gave them a longer reach and more thrust than could be achieved throwing the spear by hand. The far end of this spear-thrower contained a small hook, which inserted into the hollowed tail of a spear or dart. The hunter held the near end of both the atlatl and the spear and flung them from behind like a tennis racquet, releasing the spearpoint at the instant of peak momentum. Think of the sport of jai alai, only instead of a wicker basket flinging a rubber ball, the atlatl shot a four-foot dart tipped with a sharp stone at one hundred miles per hour.

Experiments show that spears thrown by atlatls can fly the length of a

A twelfth-century Mexican warrior with one spear loaded in an atlatl (left) and another spear in the free hand. CODEX BECKER

football field and penetrate several inches into their target. Though they ceased to be used in the Midwest around 500 CE, they continued to be a weapon of choice along the Gulf Coast for one thousand years. Spanish conquistador Hernan de Soto was attacked by indigenous people using atlatls in the American South in 1539–1540. A member of his expedition compared them to attacks in Peru, where atlatl-launched darts passed completely through the armored bodies of Spanish soldiers wearing chain mail.[31]

Birdstones appear to have been fastened to the handles of atlatls to provide a more effective grip. With a smooth stone handle added, an atlatl's main shaft could be thin and light, making it more flexible, while the hunter could maintain a secure handgrip.[32]

But birdstones may have served supernatural purposes in addition to mechanical ones. In 1000 BCE, procuring game was a hunter's most important task, since nearly all aspects of life depended on the deer, elk, bison, and other mammals that he could kill. Their flesh was eaten, their skins became clothing and shelter, their sinews bound objects together, and their horns and bones were made into tools and weapons. A spear launched from a hunter's atlatl carried the fate of his family as it flew toward a bison or deer. By attaching the effigy of a bird to his weapon, he may have invested it with the power to fly fast and straight to its target.

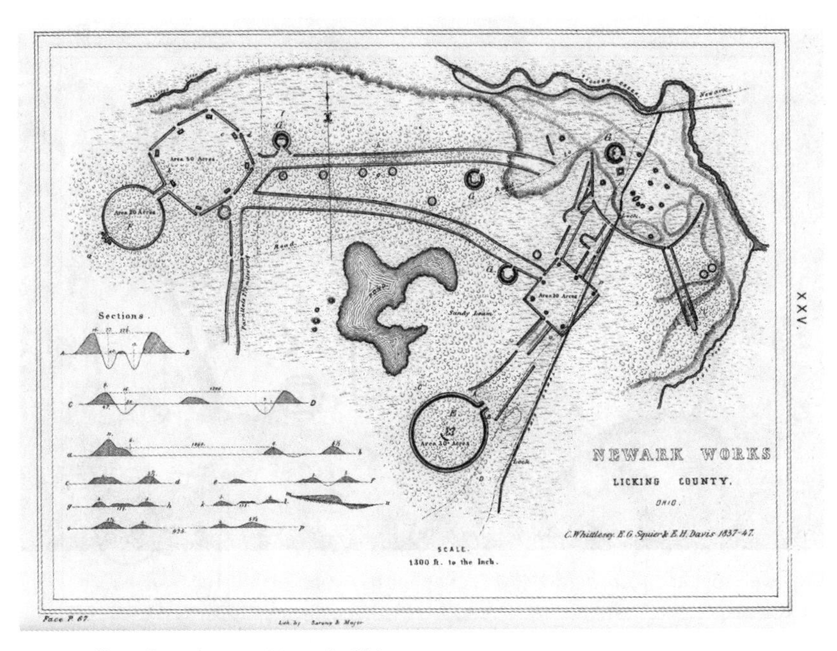

Hopewell earthworks near Newark, Ohio. FROM *ANCIENT MONUMENTS OF THE MISSISSIPPI VALLEY* BY E. G. SQUIER AND E. H. DAVIS

Birdstones gradually diminished in importance after about 200 BCE. Archaeologist Robert Hall postulates that the atlatl and birdstone evolved into long smoking pipes decorated with animal effigies. He suggests that the same visual concept may have bubbled up through the human unconscious to inspire the Hopewell platform pipe and, even later, the calumet that greeted the first European explorers.[33]

Hopewell, like Adena, refers not to a specific American Indian tribe but to a lifestyle that followed the Adena in the heart of the continent between 100 BCE and 500 CE. It is known for enormous, remarkably precise earthworks, or raised walls of soil. At the center of Hopewell culture, near modern Chillicothe, Ohio, a series of five massive, geometrically identical earthworks dominates the landscape. They are shaped in circle and square patterns enclosing ceremonial spaces that stretch for thirty miles.[34]

Like the Adena, the Hopewell inhabited a world animated by spiritual forces. Anthropologist Richard Townsend writes that they "considered themselves to be participating in a network of connections that spread outward from their communities into the life of animals and plants, leading to

Hopewell copper falcon, before 400 CE. NPS PHOTO / HOPEWELL CULTURE NHP—T. ENGBERG

the powers inherent in rivers, rocks, mountains, and other phenomena of the earth and sky, and the remote, immaterial, all powerful forces of life, death, and renewal."[35]

Their monumental earthworks shaped as squares, circles, or octagons aligned with the transits of the moon through the night sky. Inside the walls, they built temples where their shamans contacted the spirit world. Hopewell shamans often clothed themselves like the animals whose aid they sought, and they were frequently interred with owls, eagles, and hawks. Archaeologists found bodies wearing copper-reinforced headdresses of raptor feathers, effigies of birds and other animals, the claws of eagles, and even entire heads of hawks.[36]

Also buried with Hopewell shamans were sacred works of art shaped like birds. Between CE 1 and 400, the Hopewell fashioned beautiful birds from stone, copper, and mica, a mineral that separates easily into bright, mirror-like sheets. Nuggets of copper from Lake Superior were pounded flat and chiseled or cut into animal shapes as well as into tools, jewelry, and ornaments. One of the most impressive Hopewell copper birds is a falcon that would have been polished to a bright flame color and worn on the front of a shaman's headdress about two thousand years ago. Another is an eleven-inch stylized raptor's talon carefully cut out of mica. Besides

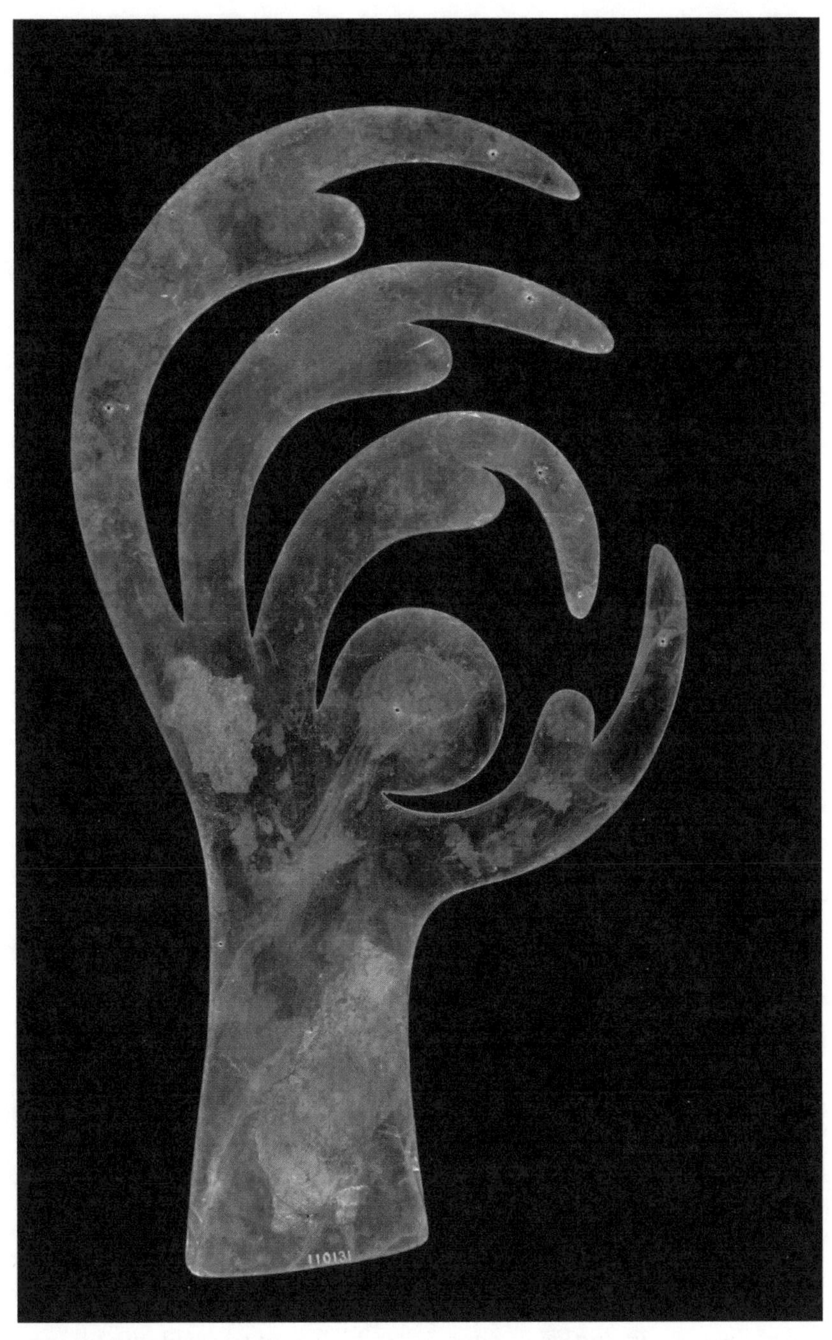

Mica raptor talon, before 400 CE. © THE FIELD MUSEUM, IMAGE NO. A110016C, CAT. NO. 110131, PHOTOGRAPHER RON TESTA

Hopewell eagle platform pipe, Naples, Illinois, before 400 CE. BROOKLYN MUSEUM, ANONYMOUS LOAN, L49.3.1. CREATIVE COMMONS-BY (PHOTO: BROOKLYN MUSEUM)

appearing in the form of copper and mica plates, bird motifs were also the most common emblems etched into pottery during the first two centuries of Hopewell life.[37]

Among a Hopewell shaman's most important tools for traveling into the spirit realms were pipes carved in the shape of animals. Smoking through pipes has been practiced since at least 1000 BCE in the Midwest. The Adena smoked with ornately decorated straight tubes, to which the Hopewell added a ninety-degree bend and a carved stone bowl. The bowl was usually located on an animal's shoulders or back, and as the smoke was inhaled it traveled directly through the heart of the effigy.[38]

Hopewell pipes may sometimes have been used for casual smoking, but their main purpose was probably religious. Archaeologist William Romain explains that to American Indians, "Breath is the animating aspect of living things. Breath is the essence of life . . . many Native American languages have a single word that means breath, life, and wind." Breath is the symbol of life; when breathing stops, so too does a creature's life. As the Hopewell shaman inhaled through an effigy pipe, smoke/breath/life passed through the animal's body, awakening its spirit, which then traveled into his own body. The souls of the shaman and animal merged.[39]

Hawks, falcons, owls, ducks, herons, cranes, and other birds all appear as Hopewell pipes, as do otters, beavers, turtles, and frogs, while some pipes are plain and undecorated. A Hopewell shaman might have used different effigies for different purposes—an owl to see in the dark, for example, or a duck to descend into the Lower World. A shaman's mystical journey was presumably facilitated by the pipe's design and the plants smoked in it.

Most Hopewell effigy pipes were constructed with the animal directly facing the shaman. "Smoking of the pipe was a mental device used by the smoker to concentrate his thoughts upon a given subject," writes anthropologist Ted Brasser. His colleague Richard Brown points out that the holes bored through effigy platform pipes were probably too narrow to admit a long cane pipe stem, so a Hopewell shaman may have been literally eyeball-to-eyeball with his spirit guide when he inhaled.[40]

The only plant residue found in these pipes was *Nicotiana rustica*, a powerful species of tobacco quite different from the commercial variety we know today. "With nicotine concentrations in excess of eighteen percent," writes anthropologist Johannes Wilbert, "cigars of this tobacco species are strong enough to produce hallucinations and catatonia." Other anthropologists found that it "has been employed to trigger ecstatic states very similar, perhaps even identical, to those induced by the 'true hallucinogens,'" including out-of-body experiences, vertigo, and hallucinations.[41] Among historic tribes, this sacred tobacco was separately cultivated and reserved for ceremonial purposes while other varieties were smoked recreationally.[42]

"According to the natives," wrote Jesuit missionary Louis Nicolas from the Great Lakes in about 1675, "this is the god of herbs and all simples; they call it Manitou Mingach . . . because the Americans find it incomparably better than their ordinary kind. They use this god to worship the other divinities that they recognize." Joseph Lafitau, who lived with the Iroquois from 1712 to 1717, noted that it was "consecrated to many uses and to various religious exercises . . . clarifying the soul, purifying it, and putting it into condition for dreams and ecstatic visions, of evoking spirits favourable to the need of the tribes who serve them."[43]

Smoking *N. rustica* may have also accompanied ingestion of much more powerful mind-altering plants. A foot-long mushroom sculpture, carved

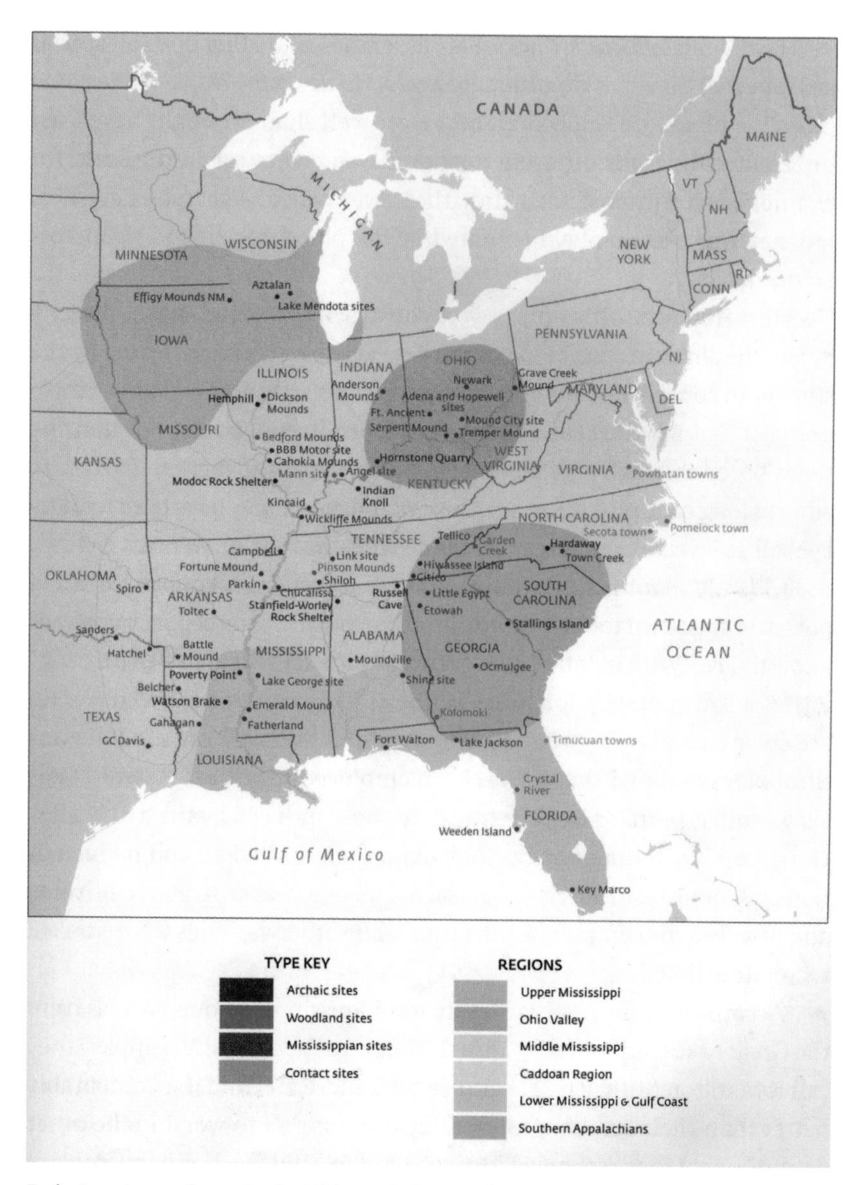

Early American cultures in the Midwest (Adena and Hopewell in green, Mississippian in yellow, and Effigy Mounds in brown). ART INSTITUTE OF CHICAGO AND MAPPING SPECIALISTS, LTD.

from wood and sheathed in copper, was laid to rest with one Hopewell shaman. This suggests that they ate psychedelic mushrooms—*Amanita muscaria, Psilocybe caerulipes, Psilocybe cyanescens, Stropharia cubensis*, or *Gymnopilus spectabilis*—that were widely available in the Midwest. Two thousand years ago, Hopewell shamans may have smoked powerful tobacco and meditated face-to-face with their bird effigy while waiting for the mushrooms to kick in.[44]

One archaeologist has even suggested that out-of-body experiences as birds could have inspired the colossal earthworks that the Hopewell created. More than 170 of these enormous enclosures are scattered across hundreds of miles in the Ohio River Valley. Many are carefully shaped in precise geometrical patterns that make sense only when seen from the air. Anthropologist Marlene Dobkin de Rios suggests that their designs might have been imagined by the shaman "from the heights of ecstasy through which he soars" and inspired by the "aerial gaze" obtained through out-of-body experiences.[45]

Similar visions may have inspired the Effigy Mound culture that followed the Hopewell in the Upper Mississippi Valley, where another people shaped birds not from stone or copper, but from the very earth itself.

Hopewell society began to wane around 400 CE. That's also when residents of the Midwest gradually replaced the atlatl with the bow and arrow, began to cultivate corn and beans, built permanent villages, and started decorating pottery with complex reflections of their beliefs. They spent more of each year in villages near their crops and less of it pursuing game from place to place. Population density rose, pushing some people out of low-lying plains onto elevated hills and ridges. Anthropologists speculate that this led to increased warfare as competing groups fought over the most fertile croplands alongside streams and rivers.

In the Mississippi Valley, some of these ancient Indians made cord-impressed clay pots with images of Upper and Lower World creatures, including birds and shamans dressed as birds. They painted similar pictures in caves and under overhanging rock ledges. They also constructed thousands of burial mounds. But instead of heaping up huge cones of dirt over several generations of their dead, as the Hopewell and Adena had done, the Effigy Mound culture sculpted low mounds over single individuals or small groups. Many of these were shaped like animals and birds.

Bird effigy mound beside State Highway 60 in Richland County, Wisconsin. MARK CUPP, LOWER WISCONSIN STATE RIVERWAY BOARD

The human remains were usually interred inside the head or the heart of the animal effigy.[46]

Beginning about 700 CE, these effigy mounds were created across Iowa, the southern half of Wisconsin, northern Illinois, and southern Minnesota. Fifteen thousand mounds were created over the next five centuries, with roughly two to three thousand sculpted to resemble birds, quadrupeds, or salamander-like creatures that archaeologists have dubbed "water spirits." They suspect that most conical mounds were made somewhat earlier than the effigies, a holdover from Hopewell traditions, and that the animal-shaped mounds reflect the transition from small band-level societies to populations with larger tribes divided into clans. The effigy animals would have corresponded with the clans.

Most effigy mounds are arranged in deliberate groups, usually located above water or near a spring. They are often positioned to create a sense of balance between Upper and Lower Worlds: wetlands may be dominated by water spirits while a bluff or hilltop may contain mostly birds. In each location a few creatures from the opposite realm are often placed, rather like each half of the Taoist yin-yang symbol containing the germ of the other in its heart, in order to balance opposing spiritual forces.

A survey of surviving effigy mounds shows that 34 percent are shaped

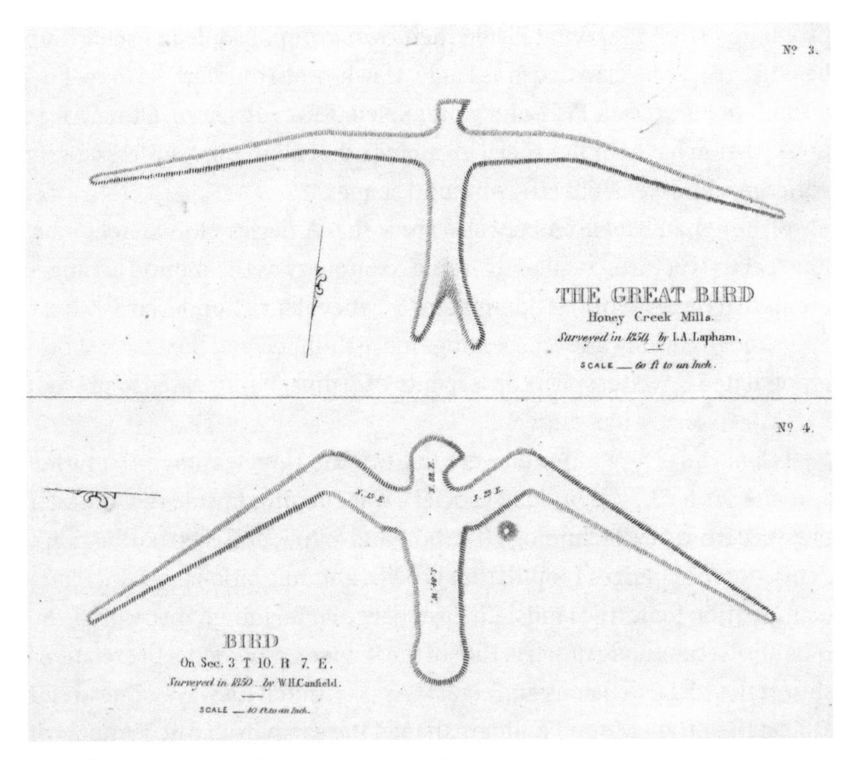

Bird effigies surveyed in Wisconsin in the 1850s. FROM *THE ANTIQUITIES OF WISCONSIN* BY INCREASE A. LAPHAM

like birds, 15 percent like mammals, and 37 percent like water spirits. Most are less than one hundred feet long and just two to four feet high. The largest bird effigy that survives is on the grounds of Mendota State Hospital in Madison, Wisconsin, with a wingspan of 624 feet. It's one of three enormous birds that guard the mouth of the Yahara River on the southern edge of a headland with many effigy mounds. Just south of the birds, this headland drops off precipitously into the deepest part of Lake Mendota, which the builders may have considered an entrance to the Lower World.[47]

After examining hundreds of effigy mounds, cave paintings, reports of digs, and ethnological studies, archaeologists Robert Birmingham and Leslie Eisenberg concluded that effigy mounds reflect a belief system similar to those of modern Ho-Chunk, Iowa, and other Siouan-speaking tribes. For example, traditional Ho-Chunk culture divides society into two large groups, "those who are above" and "those who are below," with individuals prohibited from marrying inside their own group. People in each group belong to separate clans, such as Eagle, Hawk, and Thunderbird (upper division), or Bear, Deer, Fish, and Water Spirit (lower division). Other tribes in the region have similar social groupings that align more or less closely with thousand-year-old effigy mound shapes.

Birmingham and Eisenberg also argue that American Indian religions, like social structure, parallel the same cosmology as the mound arrangements. An Upper World is dominated by benevolent Thunderbirds whose eyes shoot lightning and whose wings make thunder, and the Lower World is populated by water spirits or serpents who must be appeased to prevent harm or to secure blessings.[48]

When the effigy mounds were built in the Upper Midwest between 700 and 1100 CE, the builders' society was turning upside down under pressure from new technology (the bow and arrow, and corn cultivation), demographic changes (population growth and migrations), and warfare (competition for fertile lands). In Eisenberg and Birmingham's words, the mounds "recapitulate the structure of the universe and model the relationship of the social divisions and clans." As traditional lifeways came under attack, the Effigy Mound builders shaped the earth itself into symbols of the eternal forces that they believed would restore balance to their world, and that corresponded to their own social groups. By sculpting sacred images like the giant Mendota bird effigy into their landscape, they may have hoped to bring social life, the material world, and the spirit world into harmonious alignment.[49]

The greatest challenge to their way of life arrived in the middle of the eleventh century, just as a magnificent supernova exploded in the night sky. A new society was established 350 miles to the south that spun off satellites in every direction, including into the heart of Effigy Mound culture in Iowa and Wisconsin.

Archaeologist Tim Pauketat writes that this new society "ushered in a whole new way of life across much of the North American continent. Civilization's switch had been flipped on, and a single brilliant light instantly shown across the Midwest, South, and Eastern Plains." The new society was centered at the prehistoric metropolis of Cahokia, just across the Mississippi from modern St. Louis, and birds played a central role in its material and cultural life.[50]

3

Sun Kings, Shamans, and Thunderbirds

St. Louis loomed across the Mississippi to the west as I approached Cahokia on Interstate 55. Looking east, I could see wooded hills marking the edge of a suburban flatland. Looking south, my view of East St. Louis was blurred by smog. Careening down the exit ramp onto Collinsville Road, I cut straight across the floodplain that early French settlers called "American Bottom."

Littered today with decaying motels and auto body shops, this plain was dotted with more than two hundred pyramids and ceremonial mounds nine hundred years ago. At that time, thousands of identical thatched houses were laid out in an intentional plan, while crops stretched for miles in every direction. Unfortunately, for the last two hundred years, farmers have plowed Cahokia up, developers paved it over, and road builders carted it away in dump trucks. Interstate highways surround the site, each of their snaking exit ramps having obliterated part of the ancient city. When I finally stepped out of my car onto the former Grand Plaza of Cahokia and looked up at Monk's Mound, I felt a mixture of awe and grief.

But people approaching this spot nine hundred years ago saw a ten-story pyramid on the horizon for an entire day as they walked toward the metropolis. Coming closer, they would have seen on its uppermost terraces a suite of fine buildings guarded by a wooden palisade. There, nearer to the life-giving sun than his subjects, resided the king with his family, servants, and shamans in a compound of houses and temples. Some temples probably resembled those built by their descendants, with

The center of Cahokia, looking north, ca. 1100 CE. CAHOKIA MOUNDS STATE HISTORIC SITE, PAINTING BY WILLIAM R. ISEMINGER

carved bird effigies perched on the ridgetops: "The roof is in the form of a pavilion," wrote Antoine Le Page du Pratz in 1722 about one of the last surviving temple mounds at Natchez, Mississippi, ". . . and on the top of it are placed three wooden birds, twice as large as a goose, with their heads turned towards the east." Father Pierre François Xavier de Charlevoix, who visited the same temple in 1721, thought the sculptures were eagles.[51]

In its heyday, greater Cahokia had a total population approaching forty thousand people, and its rulers governed an area of twenty thousand square miles. The platform mound that held their compound was the largest structure in North America, as big as the Egyptian pyramids at Giza. In 1100 CE, Cahokia was one of the world's great cities, larger than London. Satellite communities laid out in its image, with platform mounds and grand plazas, dotted landscapes as far away as Florida, Georgia, Oklahoma, and Wisconsin.[52]

The city was founded in about 1050 CE, when the heavens lit up with a supernova recorded around the globe. Thousands of residents created mounds, houses, jewelry, clothing, utensils, textiles, and sculptures unlike any seen before in the Midwest. Rulers hosted seasonal feasts and staged festivals for thousands of people at a time, some of them culminating in human sacrifices. Archaeologists refer to their common culture, which spread out of the Cahokian center to dominate most of eastern North America, as "Mississippian."

It was the first highly stratified society in the Midwest, ruled by an elite class of semi-divine kings and shamans who lived on top of pyramids. The lower classes—traders, artisans, farmers, hunters, laborers—lived in neighborhoods spread out beneath the elites. Everyone in the city could always see the homes and temples of their rulers on top of the pyramid mounds, and rulers could always keep an eye on their subjects. This social structure was radically different from the comparatively egalitarian bands of hunters and gardeners who had preceded them.

Cahokia was also the first society in the Midwest based primarily on agriculture. Farmers raised massive amounts of corn in the outlying districts to feed the city, and maize was the staple of most residents' diets. The king mediated with the sun to make sure that it returned year after year to germinate the golden corn. Shamans held ceremonies to ensure that the new stalks grew tall. Thousands of people attended seasonal festivals to mark the crop's planting and harvest. And potters made huge, ornately decorated clay vessels at centralized workshops so people could transport and store the bounty.[53]

But people cannot live on corn alone, so hunters supplemented maize with mammals, fish, and birds captured in the surrounding wetlands and prairies. When archaeologist Lucretia Kelly examined eleven thousand bits of garbage that Cahokians threw out a thousand years ago, she discovered that fish accounted for half of the animal waste, deer and other mammals about 20 percent each, and birds made up about 10 percent.

Thirty different birds were harvested around Cahokia; 70 percent of them were waterfowl, with geese, mallards, and teal topping the list. This is what one would expect from the city's location along the Mississippi River, surrounded by wetlands. Upland game birds such as turkey and prairie chicken were also found, but usually in the garbage of elite households, suggesting that they were a delicacy unavailable to commoners.[54]

Bruce Smith analyzed seven other Mississippian sites at some distance from Cahokia and found that birds made up as much as 15 percent of the meat consumed, with migratory waterfowl and turkeys being the species eaten most frequently. He discovered that twenty-two species of geese and ducks had been harvested, mostly during their spring and fall migrations, with Canada geese, mallards, and black ducks being the most common. Through statistical analysis, he confirmed that the species most often hunted were those that yielded the most meat for the least effort—animals that were seasonally abundant, easily captured, and within convenient reach of settlements.[55]

Besides eating birds, high-status Mississippians made them into clothing, wore them as headdresses, used their wings and bones as tools, wove their feathers into blankets, drew and sculpted them into art, and used them in religious ceremonies. At the Mill Creek site, in Cherokee, Iowa, half the bird remains appear to be from species that weren't hunted to be eaten: 28 percent from raptors or vultures, which held special significance in Mississippian shamanism, and 21 percent from songbirds that were probably gathered for their bright feathers.[56]

Symbolic bird imagery on Mississippian pottery and rock art foreshadows that of later tribes who inherited much of their culture. Archaeologists Thomas Emerson and Timothy Pauketat found "positive links between historically recorded ritual, icons, and myths [of modern American Indians] on the one hand and Mississippian ritual features and icons on the other."[57]

On important occasions, rulers wore capes woven from multicolored feathers. These are shown on sculpted figurines such as the eleven-inch "Resting Warrior" made at Cahokia about 1100 CE (and later drilled for use as a pipe). Early Europeans recorded seeing feather capes and blankets like this from Virginia to Louisiana. In 1610, one of them described Indians wearing "mantels made both of Turkey feathers and other fowle, so prettely wrought and woven with threeds that nothing could be discerned but the feathers, which were exceeding warme and very handsome."[58]

In addition to feathered robes, Mississippian chiefs also wore headdresses that resemble those seen by the first European visitors. Some were simple clusters of feathers shaped into a circle like a crown, while others included the claws, wings, and skulls of birds. Some even included entire skins of birds, fronted by copper plates above the forehead. "The brightly polished copper head tablet," writes art historian Susan Power, "reflecting light and surrounded by an ark of radiating red or white feathers, surely must have been compared to the magnificence of rays emanating from the celestial sun." Elite Mississippian burials contained copper plates and feathers fastened to cedar splints woven into a leather cap or headband.[59]

Beyond decorative attire, archaeologists found that birds were also used to create musical instruments. Leg bones of eagles, hawks, turkeys, and cranes provided flutes or whistles for ritual use. Although the Adena and Hopewell appear not to have used bone flutes, at least fifty specimens have been found in Mississippian sites from the Ohio Valley. These are mostly cut to a length of three to five inches and perforated with a single or three holes.[60]

Resting Warrior wearing feather cape, Cahokia, ca. 1100 CE. THE UNIVERSITY OF ARKANSAS MUSEUM COLLECTIONS

Wings of large birds were also used as fans, as depicted on a Mississippian shell engraving from Spiro, Oklahoma. Piles of long wing bones, from which the primary feathers had deteriorated over the last thousand years, were often found in Mississippian graves when the rest of the bird's skeleton was missing. Wings of swans and golden and bald eagles were most often used as fans, and one site contained the long wing bones of dozens of sandhill cranes. Individual long feathers were also attached to warriors' hair, ears, and weapons.[61]

Bird wing fan held by warrior depicted on a Mississippian shell, ca. 1100 CE. OKLAHOMA ANTHROPOLOGICAL SOCIETY, *MEMOIR 1: SPIRO SHELL ENGRAVINGS* BY LATHEL DUFFIELD (1964), PLATE 9, NUMBERS 1-2

Feathers from birds of prey were so highly valued during this time that they were shipped hundreds of miles. One site in western Iowa dating to around 1100–1200 CE contained only the detached legs and talons from dozens of raptors. Archaeologist Richard Fishel suggested that their skins, wings, and feathers had been traded to Mississippian allies to the southeast: "The raptor body would have functioned as a convenient package that provided numerous items: the body itself to serve as a medicine bag, feathers for ornaments and arrow fletching, and wings for fans." The legs and claws, for which there was no demand, were removed and left behind in Iowa before the valuable parts were shipped downriver to Cahokia.[62]

None of these uses is very surprising. Mississippians, like people before and after them, probably ate everything that they could get their hands on, and used everything in their environment that would help provide clothing and shelter. What sets them apart from other societies is their strong preference for birds as religious symbols. Avian motifs are the most common designs on Mississippian ceramics, utensils, sculpture, architecture, and cave walls, appearing rather like angels do in medieval and Renaissance Christian art. Mississippian craftsmen left so much art that archaeologists can deduce the outlines of their worldview with some confidence.

To the Mississippians, the material world could sustain humanity, animals, and plants only if it remained in spiritual balance. The earth, like a spinning top, was continually in danger of falling into chaos. The Upper World promoted balance through the cosmic force of the sun and stars. The sun rose at a slightly different point on the edge of the sky every morning, lighting the world each day, warming the earth each spring, and nurturing the all-important corn. Constellations in the night sky rotated throughout the year as the dome of the heavens turned.

At Cahokia, these celestial movements were tracked by a circular calendar of tall posts that archaeologists dubbed "woodhenge." It was discovered in 1960 by Warren Wittry, who had excavated Wisconsin's Raddatz Rockshelter three years earlier. When seen by kings and shamans from the top of Monk's Mound, woodhenge's 360-degree circumference measured the movement of the sun and stars like clockwork. For many years, Wittry and his colleague Robert Hall traveled from Wisconsin to observe the equinoxes from Monk's Mound, watching the sun rise above

Cahokia's re-created woodhenge as American Indians had done one thousand years before.[63]

The ruler of Cahokia, who resided on top of Monk's Mound, closest to the source of all life, interceded between the worlds. He wore burnished copper ornaments that shot the sun's light into the eyes of his subjects, and he was usually depicted on pottery or rock art with solar rays emanating from his head. By carefully performing seasonal rituals, he tried to align human affairs with cosmic ones. When life went out of balance—in times of drought, pestilence, tornadoes, long winters—his shamans entered trances in which they left their bodies to become birds who intervened in the Upper World. They are frequently depicted as "birdmen" on Mississippian copper plates, cave walls, clay tablets, stone sculptures, and shell cups.

"Despite the diversity of images and symbols connected with the upper world," writes Susan Power, "no other surpasses the recurrent and primary use of birds (and parts of them) as dominant symbolic links with the upper world." This included not only actual physical birds, but also unseen Thunderbirds whose eyes shot lightning, whose wings made thunder, and who were considered benevolent toward humans.[64]

The Lower World, in contrast, was the source of chaos and imbalance. It connected to the human realm through springs that bubbled up from underground, through the depths of lakes and rivers, and from inside the deep caves found throughout the central Mississippi Valley. The Lower World was symbolized in art by serpents who came out of holes in the ground each spring, salamanders like the giant aquatic amphibian hellbender *Cryptobranchus alleganiensis*, and a horned water monster, which could have been inspired by enormous Mississippi River catfish or alligator gars, that overturned canoes.

The Lower World was often symbolized as a horned dragon or panther in Mississippian art. These subterranean and underwater spirits were perpetually at war with the Thunderbirds and were considered hostile toward humans. They had to be carefully placated and appeased.

The *axis mundi*, or Tree of Life, penetrated through all three realms (Upper, Lower, and terrestrial) and facilitated the movement of spirits and shamans through the universe. Each Mississippian village had a tall ceremonial post erected prominently in the center of its main plaza to symbolize the *axis mundi* and connect the three realms. In some communities this was complemented by a circular woodhenge like Cahokia's that divided the world horizontally into four cardinal directions and 360 degrees of celestial movement around the horizon. The Upper, Lower, and human realms were thus reflected in the geography of each Mississippian town: the ceremonial axis in the center pointed to the heavens and to the netherworld while the clockwork woodhenge measured daily and seasonal cycles in this world.

The Mississippian three-tiered cosmos was literally carved in rock at Millstone Bluff, in southeastern Illinois. About 1300 CE, residents on this hilltop village scraped three groups of images into a hundred-meter-long sandstone slab on the shoulder of the bluff, just below their homes. The eastern panel (Upper World, facing the rising sun) contains a falcon, a bird-man, chevrons, and other celestial motifs, while the western one (Lower World) depicts snakes, a horned serpent, and subterranean beings. The central panel contains elements of the other two as well as a cross-in-circle image that shows the Tree of Life cutting vertically through the cosmos, the four cardinal directions quartering the world, and its circular horizon. This carved sandstone map of the universe would have been conspicuous to all Millstone Bluff residents, since their path to the bluff's only water source ran directly above it. Like a medieval cathedral or a Puritan church atop a hill, it was a constant reminder of the Mississippians' place in the world.[65]

Because to the Mississippians all of nature was infused with spiritual power, human activity on the material plane was implicitly religious. There was no meaningful separation between secular and sacred. The very landscape itself expressed spiritual power: every Mississippian village, from its highest pyramid and central *axis mundi* to its flat plaza and watery lowlands, expressed the perpetual balancing act between Upper and Lower Worlds, between harmony and chaos.

Images pressed on pots, carved into shells, embossed on copper ritual objects, modeled into earrings and ornaments, painted on cave walls, and cut into bluffs all reinforced the sacredness of daily life. Every human action either encouraged or impeded spiritual harmony. Sun Kings, shamans, farmers, hunters, warriors, and laborers all had a place in the cosmic balancing act. True harmony could be achieved only by embracing both sides of every duality: sky and earth, Thunderbirds and Water Panthers, matter and spirit, order and chaos.[66]

Bird images expressing this tension appear on thousands of pots, cave walls, sculptures, pipes, clothing, and ritual objects. The species most frequently displayed are peregrine falcons, eagles, owls, and woodpeckers, though ducks and other birds also frequently appear. Sometimes the images are explicit and lifelike; other times, abstract icons symbolize the Upper World. These avian motifs typically represent the forces of the Upper World in its perennial conflict with the chaos and darkness of the Lower World.

Often a single diagnostic feature of a bird was used to stand for the whole. For example, the peregrine falcon's unique crescent-shaped eye surround morphed into an abstract "forked-eye" or "weeping-eye" motif that appears alone in many contexts to convey the falcon's speed and ferocity. Abstract feather or chevron designs were often drawn side-by-side like twining vines (called "petaloid" borders by archaeologists) to signify that the subjects or events they surrounded were located in the Upper World.[67]

The large pots in which Mississippians stored their corn contain a repertoire of bird motifs. These ceramic vessels have wide shoulders and broad, inward-leaning lips decorated with symbols of the Upper World. The more realistic images on them include the falcon's forked-eye pattern, fan-shaped tails, and solar circles. Other abstract Upper World patterns on the corn vessels include feathers, zigzag lightning bolts, arcs, and chevrons.[68]

Ritual objects used by shamans usually contain bird imagery. Hundreds of marine shells imported from the Gulf Coast to Mississippian sites farther north are decorated with incised drawings of feathers and avian motifs. They are thought to have been used as vessels from which ritual, perhaps psychoactive, beverages were drunk. Mississippian artists also cut, pounded, engraved, and embossed copper plates with falcon-inspired forked eyes and other avian motifs. Textile fragments, possibly part of a shaman's cloak worn during ceremonies, were found printed with a barred-wing raptor and a horned water spirit, representing both sides of the cosmic balance.[69]

No bird was more important in Mississippian iconography than the falcon. "There should be no surprise," writes archaeologist James Brown, who excavated the Mississippian ceremonial center at Spiro, Oklahoma, "in finding the falcon occupying such an important role in Spiro symbol-

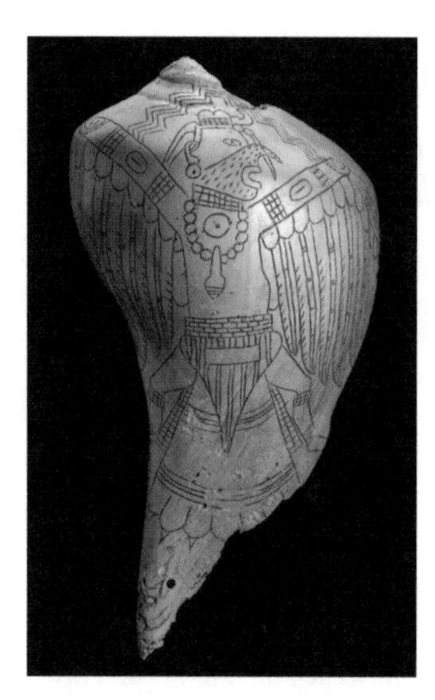

(Above left) Incised whelk shell cup showing falcon-warrior from Spiro, Oklahoma, ca. 1100 CE. NATIONAL MUSEUM OF THE AMERICAN INDIAN, SMITHSONIAN INSTITUTION (18/9121). PHOTO BY NMAI PHOTO SERVICES. ***(Above right)*** Drawing of the design on the whelk shell at left. FROM *PRE-COLUMBIAN SHELL ENGRAVINGS FROM THE CRAIG MOUND AT SPIRO, OKLAHOMA* BY PHILIP PHILLIPS

ism, since the falcon is famous for its spectacular aerial attacks that literally knock ducks and smaller birds out of the air."[70]

A human figure transformed into a falcon is a common image on Mississippian artifacts. This winged being has often been called a "falcon-warrior," and was embossed in copper, engraved in shell, modeled in clay, and sculpted in stone. A falcon-warrior ensemble of clothing and ritual accessories was even found at Spiro, Oklahoma, near an incised shell showing a wonderful version of the image.[71]

The falcon costume and image suggest that the shell may depict a shaman who has transformed into a falcon to visit the Upper World, rather than merely a powerful warrior. Some archaeologists identify this deity with the Morning Star (Venus) that rises into the sky before dawn and pulls up the sun. They suggest that the Falcon-Warrior/Morning Star motif symbolized creation, rebirth, and the restoration of balance at the start of each day. It exists in many media, including copper and clay tablets, and persisted into modern times among tribes thought to have descended from the Cahokians.[72]

(Above left) Copper embossed falcon-warrior, Cahokia, ca. 1200–1400 CE. MILDRED LANE KEMPER ART MUSEUM, WASHINGTON UNIVERSITY IN ST. LOUIS. GIFT OF J. MAX WULFING, 1937. ***(Above right)*** Sandstone engraved falcon-warrior, Cahokia, ca. 1300 CE. CAHOKIA MOUNDS STATE HISTORIC SITE

Mississippian shamans smoked tobacco and other herbs, like their Hopewell predecessors, but bird effigy pipes are quite rare. Mississippians had access to the same hallucinogens used by their predecessors, including cultivated *Nicotiana rustica*, psychedelic mushrooms, sweet flag (*Acorus calamus*), jimsonweed (*Datura stramonium*), and morning glory seeds (*Ipomea violacea*). However, the latter two plants were prepared by historic American Indian nations as beverages rather than smoked, and Mississippians may have drunk from shell cups like the one shown on the opposite page. At least one effigy pipe from this culture is in the form of a duck and several are frogs, suggesting they may have been used as vehicles to the Lower World.[73]

Birds and birdmen were also carved or painted onto cave walls all around the Mississippian world, including subterranean sites in Missouri, Tennessee, and Wisconsin. Some were etched (petroglyphs) beneath overhanging ledges in exposed locations, perhaps intended to be seen by people passing beneath. Others were painted deep inside dark caves that were probably used for rituals and ceremonies and meant to be seen only by initiates holding torches. In both settings, writes archaeologist Carol Diaz-Granados, "The bird is the most frequently depicted icon in the rock art both north and south of the Missouri River, with the birdman theme evident in the selection of these depictions."[74]

This might have been a deliberate attempt to balance Lower World forces in subterranean spaces by introducing symbols of the Upper World. Some of the images in remote caves closely imitate those drawn in Cahokia and other large Mississippian centers, but others depart stylistically from them. For example, a number of underground birdmen in Missouri caves have bifurcated tails reminiscent of the swallow-tailed kite. Some of these are so deeply notched that the long tail feathers mimic legs and the figures resemble human torsos with bird-like heads. In many caves, simpler symbols such as the forked eye or petaloid feather motif were intended to evoke the Upper World.[75]

Deep caves in Tennessee contain images of birdmen, abstract symbols, and realistic depictions of horned owls and pileated or ivory-billed woodpeckers. These are painted on walls above charcoal residue from torches that has been radiocarbon-dated to 1200 CE. Archaeologists think these caves were sanctuaries where shamans connected with spirits of the Lower World, or locations where boys went through spiritual initiation under the guidance of older men. Paleofecal remains in the caves contained residue from the plant sweet flag (*Acorus calamus*), suggesting that the people who used the caves consumed at least one hallucinogenic plant. Entoptic designs—abstract geometric forms typically found in shamanic art around the world and frequently seen by users of psychedelic drugs such as LSD and DMT—are also carved into or painted on these Tennessee cave walls.[76]

Other Mississippian cave pictographs illustrate a narrative still told aloud today by the Ho-Chunk and Iowa tribes. These are painted on the walls of Picture Cave, sixty miles up the Missouri River from Cahokia, and in four other Missouri sites. They were also painted deep inside the Gottschall Rockshelter in southwestern Wisconsin, which contains forty pictographs arranged in six panels that have been radiocarbon-dated to 900–1000 CE.[77]

Archaeologists believe these images illustrate episodes from a cycle of stories about the hero Red Horn. Some even suspect that all the falcon-warrior images discussed earlier—the Resting Warrior statue with the feathered mantle, the Cahokia birdman clay tablet, the Spiro incised shells, and the Missouri copper tablets—are representations of Red Horn.[78] A very simplified version of his story, based on a version told by a Ho-Chunk elder to anthropologist Paul Radin one hundred years ago, goes like this:

Red Horn starts life resembling something similar to the Ugly Duckling or Cinderella figures of European fables. The youngest of ten brothers, he is disparaged and called He-Who-Gets-Hit-with-Deer-Lungs after the others throw garbage at him. But he astonishes them when he competes in a race around the rim of the world and wins by turning himself into an arrow (a falcon symbol) and shooting past all the other contestants, including several bird spirits that have taken material form in order to compete.

That's when he reveals that he is not really a mortal at all but a spirit, and that in the Upper World his name is He-Who-Wears-Human-Heads-in-His-Ears. Small human heads then magically appear dangling from his earlobes, and he strokes from his skull a long braid of red hair, telling his brothers that they should call him Red Horn from now on. Although the prize for winning the race is the beautiful daughter of a chief, he chooses instead for his first wife an orphan girl who, like him, had been belittled and despised.

In a series of subsequent episodes, Red Horn makes friends with a magical turtle, a Thunderbird named Storms-as-He-Walks, and other animal spirits who have incarnated in the human world. Because he sometimes flies into the heavens to accomplish tasks, he is always depicted in rock art and on pottery with the peregrine falcon's crescent eye surround and other avian motifs.

When a tribe of giants attacks his people, Red Horn and his friends defeat them in a string of athletic contests; Storms-as-He-Walks stuns his enemies by striking them with claps of thunder. At Gottschall, the giants are depicted with sunray headdresses and beaded forelocks, probably symbols of the Cahokian elite. At the end of the contests, all the giants are put to death except a female chief, Red Horn's fiercest adversary, whom he takes as his second wife. Both of Red Horn's wives eventually bear sons.

This drawing of the main Red Horn panel from the Gottschall Rockshelter in Southwest Wisconsin was adapted from black-and-white tracings of early photographs by Mary Steinhauer. ROBERT BOSZHARDT

But the tables are turned when more giants attack. This time, Red Horn and his companions are killed and the invading giants wipe out his people. His two sons escape the slaughter, however, and sneak into the giants' village. There they turn themselves into feathers, float up to the top of the Tree of Life (*axis mundi*), recover the scalps of Red Horn and his friends, and escape. As the giants chase them, the boys' quivers are kept magically full of arrows (more falcon imagery) no matter how many times they shoot at their pursuers. After reaching home, the sons miraculously resurrect Red Horn, Turtle, Storms-as-He-Walks, and the murdered villagers. At the request of Storms-as-He-Walks, the Thunderbirds donate sacred weapons to the people so they can always protect themselves in the future.[79]

Even this short summary (the full cycle of tales is much longer and richer) demonstrates that the Red Horn narratives are dominated by bird motifs. The protagonist, Red Horn, is a classic falcon-warrior, able to assume human or avian shape at will. His raptor-like strength and speed make him a formidable warrior. His Thunderbird ally fights alongside him and gives weapons to his people. His sons become feathers, which rise to the top of the *axis mundi* and resurrect him. Because the pictorial versions on cave walls have been carbon-dated, we know that these stories about a magical birdman have been carefully told and retold in the heart of America for a thousand years.

Red Horn's son carrying his father's head away from the village of the giants; he is wearing long-nosed god/human-head earrings, 1200 CE. Picture Cave, Missouri. THE UNIVERSITY OF ALABAMA PRESS

The link between the cycle of stories about Red Horn and the pictographs is not always clear. The images sometimes show extraneous characters not mentioned in the oral tales. In one cave, Red Horn is androgynous or female; some archaeologists believe the image there may represent one of his wives. The art does not always follow the tales as closely as images in a graphic novel follow its text. Instead, the pictographs may have served as selective reminders of specific episodes, like markers or signs, rather than forming an uninterrupted visual narrative.[80]

Some archaeologists speculate that these stories and images may document actual historical events. At the time the Gottschall images were painted in Wisconsin, about 950 CE, the Effigy Mound people of the Upper Mississippi Valley were under pressure from the new Cahokian society impinging on them from the south. Mississippian fortified towns with platform mounds were being erected at Trempealeau and Aztalan, Wisconsin, and at the latter site there is abundant physical evidence of violent conflict between two different peoples.

Did the first attack of the giants in the Red Horn stories represent the incursion of Mississippians into Effigy Mound territory? Were the invad-

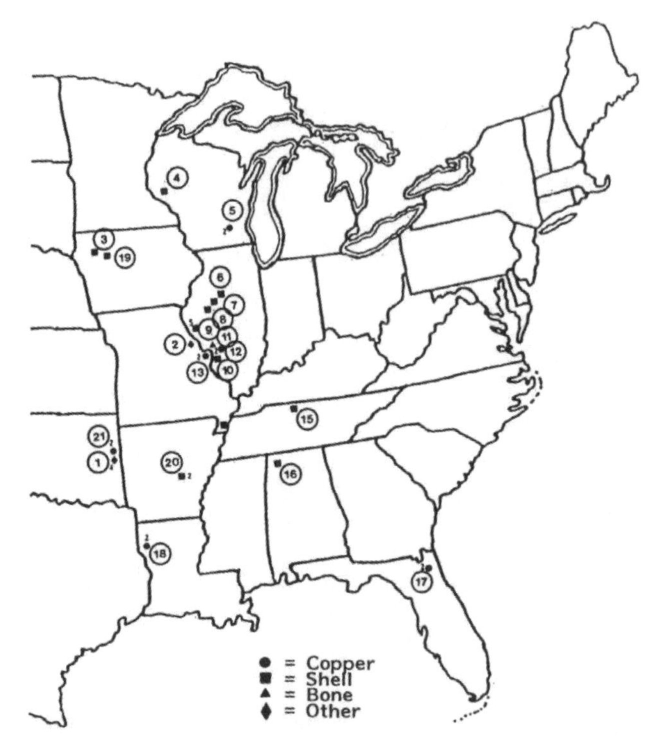

This map shows the distribution of long-nosed maskettes in bone, shell, copper, or other media. THE UNIVERSITY OF ALABAMA PRESS

ers initially defeated, only to triumph in a later invasion? Did the survivors (Red Horn's descendants) retreat to a safe distance and try to preserve their own culture? To some archaeologists, the paintings hidden deep in caves and the Red Horn cycle that was orally communicated for ten centuries may recount Cahokian invaders entering the northern homelands of the Effigy Mound builders a thousand years ago.[81]

Red Horn, in his character of He-Who-Wears-Human-Heads-in-His-Ears, became a widespread icon throughout the entire Mississippian world. Actual earrings in the shape of human faces, sometimes the face of the "long-nosed god," as well as depictions of people wearing them, have been found in archaeological sites all across the Midwest.

Just as the forked-eye motif was used to symbolize the fierceness of falcons, the human-head earrings epitomized the falcon-warrior Red Horn and evoked the narratives about his exploits.[82]

Earrings worn on faces; stories told for generations; images carved into ceramic pots, copper plates, and shell cups; and pictographs painted by torchlight all demonstrate the central place that birds held in Mississippian culture. We know that people of the Adena and Hopewell cultures used birds in their religious rituals in the Ohio Valley centuries earlier; their shamans were buried with avian-inspired art and ceremonial objects. But birds permeated all aspects of the later Mississippian culture, appearing not only as religious icons but also on everyday objects like cups and bowls, in stories told aloud, and even sculpted into the land itself as at Millstone Bluff. And birds played this central role not just in a narrow geographic area like the Ohio Valley but throughout tens of thousands of square miles in the Midwest.

Forked-eye motif on a sandstone maskette found in La Crosse County, Wisconsin. UNIVERSITY OF WISCONSIN–LA CROSSE

Archaeologist James Brown meticulously teased out meanings from this visual imagery and the Red Horn cycle. He summarized them as illustrating "a contest between the forces of life and those of death, in which there is no clear-cut winner, only an alternation of winners in a continuous cycle of defeat followed by victory." The Red Horn narratives and birdman imagery validated the most basic tenets of early American religion: the interdependence of Upper, human, and Lower realms, the movement of spirits through all three, the eternal tension between balance and chaos (the unity of opposites), and the perpetual cycle of death, renewal, and rebirth.[83]

After roughly two hundred years, Cahokia faded as the center of Mississippian life for reasons that are not yet well understood. Its cultural energy shifted to the south and southeast, where a few generations later Sun Kings, birdmen, and platform mounds greeted the first Europeans. A few of these towns, like Natchez, survived as late as the eighteenth century. By then, the former metropolis of Cahokia had been abandoned, its inhabitants scattered throughout the Midwest and onto the Plains. American Bottom, an area in the floodplain of the Mississippi across from modern St. Louis, became a deserted backwater.

Modern Indian nations with the closest cultural connections to Cahokia include the Pawnee, Osage, Kansa, Ponca, Omaha, Quapaw, Iowa, Otoe, Missouri, Ho-Chunk, Mandan, Hidatsa, and Crow. Although their ancient capital disappeared, the culture of the Mississippians has continued to inspire people across the continent's heartland until our own time.[84]

North of Cahokia, however, a new society arose between 1200 and 1600 CE for whom birds were also central symbols. Just as archaeologists aren't sure what became of the Cahokians, they don't know precisely where or when the Oneota originated, or precisely how they may have been related to Mississippians. The two cultures overlapped in some places and exchanged goods, and the Oneota blended many features of Mississippian culture with original practices of their own.

The Oneota concentrated in villages without platform mounds or semi-divine rulers across Wisconsin, Illinois, Iowa, Missouri, Minnesota, and the eastern Great Plains. Families and clans were the principal social and economic units, as opposed to the multitiered Mississippian society in which a king ruled over specialized classes and controlled production of food and commodities. Oneota towns were semipermanent, occupied just for the spring planting and fall harvesting seasons. During the summer and winter, extended families dispersed in bands to travel the large open spaces to capture bison and deer.

Although they cultivated corn and vegetables in carefully ridged fields, the Oneota didn't rely on agriculture to the same degree that the Mississippians had. They lived in large oval longhouses or key-shaped smaller ones and preserved their corn in underground storage pits, to be eaten when they returned from hunting. Scholars generally agree that they were the immediate ancestors of the modern tribes who were living in the Upper Midwest when Europeans arrived in the seventeenth century.[85]

Although the Oneota didn't duplicate Cahokia's political or economic systems, they retained much of its cultural life, including the symbolic use of birds. The oral recounting of Red Horn's story and the visual motifs related to it, for example, were passed down over the centuries through the Oneota to modern tribes. And, like all the indigenous peoples before and after them, they ate birds and used them for utilitarian purposes.

The bones of at least thirty-nine bird species have been found in the refuse pits of Oneota houses, with waterfowl being the most numerous. Passenger pigeons, a dozen species of ducks, trumpeter swans, and Canada

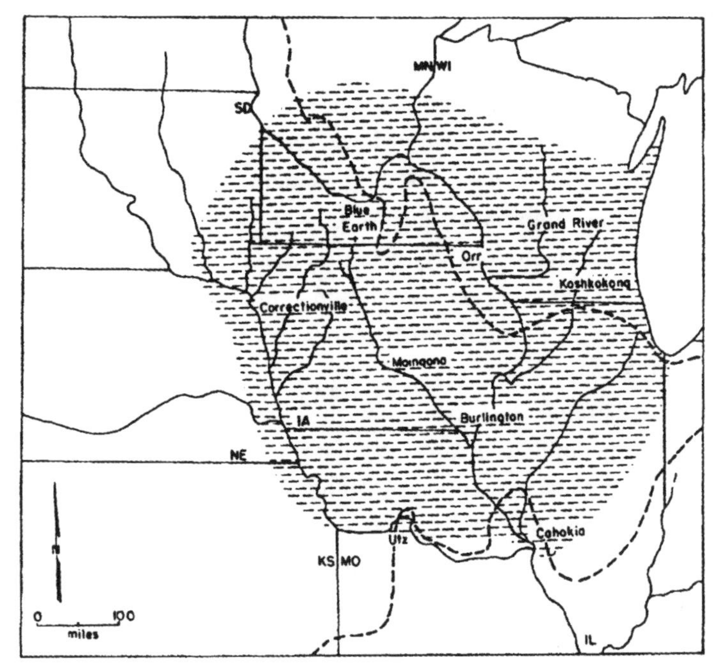

Distribution of Oneota villages and culture, ca. 1200–1600. FROM "HAWKS, SERPENTS, AND BIRD-MEN" BY D. W. BENN

geese were eaten, along with their eggs. Turkey bones are widely dispersed across the Oneota region, including wing bones. Archaeologists suspect that Oneota people may have preferred turkey feathers for fletching arrows, in addition to desiring turkey as a food source. As in earlier times, birds had a minimal effect on Oneota diets, providing less than 10 percent of total calories at every site where they were found.

Birds that were taken for other reasons than food included hawks, falcons, eagles, sandhill cranes, orioles, cardinals, and red-winged blackbirds, all of which would have provided feathers or other body parts for decorative, ceremonial, or practical uses. A few charred bones of red-winged blackbirds suggest that this ubiquitous species may sometimes have been cooked and eaten.[86]

The Oneota also used birds in religious settings. In two burials discovered in Allamakee County, Iowa, the beak of a raven and the front portion of a whooping crane's skull were laid next to a human's head. A headdress excavated from a grave in Sauk County, Wisconsin, included two similarly amputated raven skulls, as well as bones of the ravens' four wings; all six pieces had probably been attached to a headband. The beak of an ivory-billed woodpecker, possibly drilled for attachment to a headdress and presumably acquired through downriver trade, was unearthed in Rock Island County, Illinois. Just west of the Oneota in central South Dakota, the remains in contemporaneous Arikara burial sites included the skulls of ducks, crows, ravens, hawks, owls, and eagles strategically placed next to the deceased. Birds of prey were found alongside adult men, crows and ravens with children, and no birds at all were buried with Arikara infants or women.

Although archaeologists meticulously described these objects, they have been reluctant to speculate on what the birds may have meant to the Oneota or why there were apparent gender divisions in their use; we can safely assume that birds functioned as symbols of the Upper World, but we can't claim to know what individual species may have represented in these burials. [87]

The Oneota conspicuously displayed two avian motifs derived from Mississippian models. Archaeologist David Benn notes that the two clans who lead government and politics in today's Ho-Chunk society, the Thunderbird and Warrior clans, also dominated Oneota society. He argues that these two clans employed Thunderbird and falcon-warrior images to reinforce their hegemony, rather like the present-day US military displays the stars and stripes.[88]

The most common motif used on Oneota pottery was the chevron. This is an abstract, idealized representation of the wing and tail shapes of raptors and was a symbol of the Thunderbird (governing) clan. The chevron was also found in Thunderbirds depicted in rock art on cliffs and in caves, on stone tablets, and in other media. Based on analogy with historic tribes, archaeologist Thomas Berres believes that Oneota women made the tribe's ceramic vessels and decorated them with these abstract bird motifs, and that the woven bags and baskets made by women also displayed the abstract zigzag, triangle, and hourglass symbols of the Thunderbird.[89]

Equally common among the Oneota was the birdman or falcon-warrior motif. "The hawk symbolism is so pervasive in Oneota representations," Benn writes, "that it must be presumed the warrior theme is being expressed. The Oneota falcon-impersonator was an idealized human being;

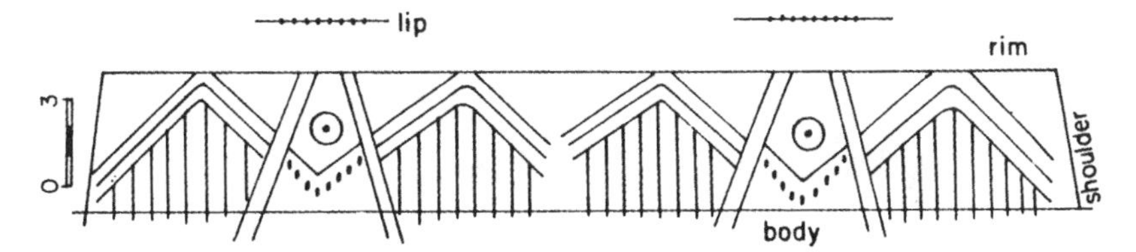

Chevron design on an Oneota ceramic vessel from Iowa. FROM "HAWKS, SERPENTS, AND BIRD-MEN" BY D. W. BENN

a model for warriors who inherited social advantages through descent and had achieved positions of leadership through exploits in war and hunting." Benn writes from the perspective of Marxist historical materialism and discounts any purely religious significance for these images in favor of economic and political explanations. A more religiously inclined observer, however, might conclude that the forked eye, the chevrons, and the winged birdmen functioned as sacred icons, as they had for the Mississippians, reminding Oneota people who saw them of their place in the cosmos.[90]

In addition to decorating pottery and ritual objects, the Oneota painted or carved images of birds directly into the landscape that surrounded them—inside caves, on boulders and rock shelter walls, on ledges and outcrops, and above riverbanks. Petroglyphs (incised stone bas-reliefs) and pictographs (painted surface images) in the eastern half of the United States are concentrated in the Oneota region of the Upper Mississippi Valley. There are 50 rock art sites in Illinois, 116 in Arkansas (mostly in the north), 17 in Iowa, 134 in Missouri, and more than 100 in Wisconsin. At the Jeffers Site in southwestern Minnesota, more than four thousand images, including representational and abstract birds, were carved into a rock ridge fifty yards wide and three hundred yards long that protrudes slightly above the surrounding prairie. Dating rock art is difficult for a variety of reasons, but scholars tend to agree that the majority was made during the late Mississippian and Oneota periods.[91]

A number of repeated motifs appear in the rock art of the Upper Mississippi that correspond roughly to images on later Ojibwe and Menominee birch bark scrolls, suggesting that they functioned rather like hieroglyphic texts. "To those native Americans with esoteric knowledge concerning their culture's worldview," writes archaeologist Thomas Berres, "the figures in picture writing were invaluable memory aids that were passed

Realistic bird shapes found on Rocky Hollow, Missouri, petroglyphs. THE UNIVERSITY OF ALABAMA PRESS

down through successive generations." Archaeologist Robert Salzer considers them a sort of visual encyclopedia of religious concepts that were publicly proclaimed by shamans. Rock art sites were sacred locations where ceremonies were held, human remains might be interred, and rituals were conducted. The most common design elements found in Oneota rock art from Missouri to Minnesota are avian images, suggesting that birds remained centrally important to indigenous cultural life as they were represented in the newer media of pictographs and petroglyphs.[92]

Many avian images look like actual birds, while others are idealized or abstracted into nested chevrons and semicircles symbolizing avian themes. These images may have represented clans, prey that was hunted, an individual's spirit guide (see next chapter), or a respected animal adversary. Thunderbird images found all across the region could represent living birds, creatures of the Upper World, ancestral spirits of the Thunderbird clan, or shamans during trance. At the confluence of the Missouri and Mississippi Rivers—a location where two mighty forces from the Lower World join together—the Thunderbird, a powerful symbol of the Upper World, is the most widespread and conspicuous motif depicted on the bluffs, suggesting an attempt to balance cosmic forces by invoking heavenly spirits above the waters rushing toward the underworld.[93]

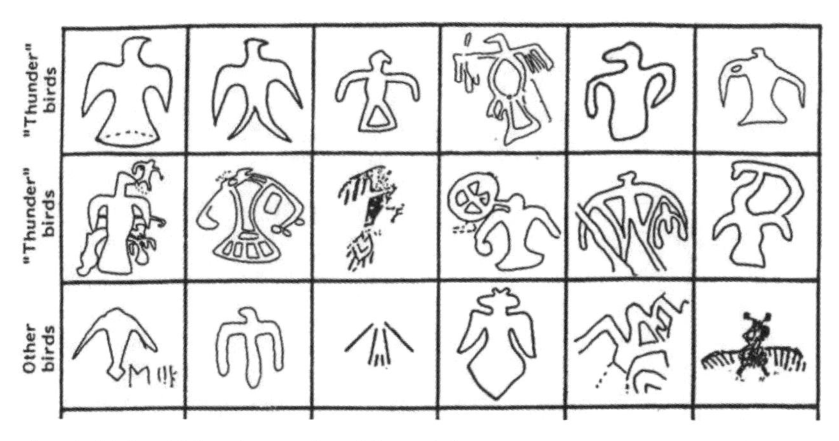

Thunderbird and falcon images from Missouri sites. THE UNIVERSITY OF ALABAMA PRESS

Birds were not so important, however, to all indigenous peoples. East of the Oneota in Indiana, Ohio, and upriver into Tennessee and West Virginia, a culture that archaeologists call Fort Ancient also adapted Mississippian ways but didn't embrace birds as cultural symbols like the Oneota did. Fort Ancient people ate birds, especially turkeys, and fashioned their bones into beads, awls, pendants, and flutes, but they didn't build an iconography or, apparently, a religion around birdmen, falcons, and Thunderbirds.

The forked-eye motif occurs only occasionally on Fort Ancient ornaments. One instance of a Thunderbird occurs on a stone pipe, which may have been imported from the Oneota, and the chevron design appears sporadically on their pottery. They made bowls with ducks and other birds sculpted into their handles, and a few bird effigy pipes have been found. But apart from these examples, avian imagery is rare on Fort Ancient artifacts. They traded occasionally with the Oneota and presumably saw avian images on imported goods, but birds were clearly less important to them than to their neighbors farther north and west.[94]

4

The Great Spirit Had Blessed Them

Can archaeology really take us inside the minds of ancient people who left no written records? It can show how they acted toward birds, but can it reveal what they thought about them? No matter how long I pored over the photographs of pottery and petroglyphs or studied the reports of archaeologists, I somehow felt as if I never got much further than Alan Watts's children did with the grapes, slicing them over and over, but finding only more outsides. How can we get inside the heads and hearts of people who lived here two thousand years ago?

British anthropologist Mike Williams may have gone further down this path than any other scholar. "The stimulus for me to explore the beliefs of the past," Dr. Williams writes in his 2010 book, *Prehistoric Belief: Shamans, Trance and the Afterlife*, "stems from my participation with modern Druid orders and my practice of shamanism."

Before dismissing him as biased, remember that most writers on the history of Christianity have been practicing Christians and most historians of science are trained scientists. Using one's own beliefs to illuminate earlier ones is nothing unusual, and Williams defends that approach: "Without an understanding of the entire bounds of existence for the shaman, including the reality of the otherworlds (and possibly even experiencing it oneself), there is insufficient content for an observer to make sense of very much at all." In effect, he says we can't understand how ancient peoples thought without somehow experiencing their mystical visions ourselves.[95]

But this approach reminds me of a paradox I encountered many years ago during one of my first LSD trips. My brother and I were exploring the Harvard campus at dusk, several hours after ingesting two doses of what was called "windowpane." Beneath a shrub we noticed a cluster of mushrooms that resembled something from *Alice's Adventures in Wonderland*, undulating and pulsating and glowing with colors. We glanced at each other without speaking. Did they really exist, or were they just a lysergic hallucination? There seemed to be only one way to find out.

So, while well-dressed couples strolled past on their way to clubs and restaurants, we knelt down gingerly in the dirt and began to creep on hands and knees close enough to touch the fantastic fungi. But as we reached out our hands, we glanced silently at each other again and dissolved in laughter, simultaneously realizing that our psychedelicized fingertips would yield no more trustworthy evidence than our eyes had. The existence of the mushrooms couldn't be decided subjectively. Maybe they were real; maybe they weren't. We could never know.

So while Williams is certainly right that trying to duplicate ancient mindsets can help us make educated guesses, there's still no way to determine whether those guesses are accurate or not. Is my transcendental vision similar to a Hopewell shaman's? Maybe; maybe not. The question can't be answered experientially. To truly grasp the thoughts of vanished people, we need their words. And words about Midwestern peoples and birds began to appear on paper only when European invaders entered the interior in 1539 CE.

Those words, of course, come down to us through European filters. Much of what explorers and missionaries witnessed in Indian villages was incomprehensible to them, in the same way that the European worldview made no sense to their hosts. "We've been trying to teach you how to live here for four hundred years," an Oneida matriarch said to me once, "but you just don't get it." Her remark is a reminder that, although this chapter spans centuries, the indigenous beliefs about birds quoted here are still widely held by American Indians.

Cahokia collapsed about 1300 CE, and over the next 250 years, descendants of Mississippian, Oneota, Fort Ancient, and other indigenous peoples fanned out through the Midwest. Within a few generations, forty different American Indian nations were living across hundreds of thousands of square miles from Niagara Falls to the Dakotas in the north and Kentucky to Oklahoma in the south, each with its own unique language and culture. Despite their diversity, one of the things these peoples shared was that almost every aspect of their economic, social, and religious lives involved birds. Birds were hunted, eaten, fabricated, worn, and worshipped all across America's heartland.

The ways that American Indians interacted with birds could fill volumes, so I've limited the following discussion to three topics—Native religious beliefs about birds, the use of feathers in clothing and ritual, and methods of hunting, cooking, and eating birds—as examples of how these different cultures understood and related to birds. Whenever possible, I've relied on nineteenth- and twentieth-century Indian autobiographies to supplement early Europeans' observations and modern ethnographical studies.

All tribal peoples in the Midwest were faithful to the shamanism of their ancestors, just as Renaissance Europeans were faithful to their centuries-old Christianity. This is not to say that every member of every tribe believed that every bird was holy. As in all cultures, some people were probably nonreligious or lapsed practitioners, or nurtured different beliefs than those of the majority. Beginning in the seventeenth century there were even Christian Indians.

But the underlying beliefs of American Indians throughout the Midwest in 1850 were essentially the same as they had been for thousands of years: the entire universe is sacred; the human world is permeated by supernatural forces that become visible during altered states of consciousness; and the Upper, Lower, and material realms must be kept in balance. Thousands of American Indians (and other people around the globe) continue to embrace this worldview today.

It was accepted as common sense in tribal communities that, in the words of modern Algonquian writer Evan Pritchard, "Where you are standing is sacred ground and every day is the sabbath." Lutheran minister Reverend Thomas Mails lived with various Plains tribes in the 1950s and 1960s and wrote that "in their quest to reach an accord with all things, the Indians assumed that invisible presences existed to answer every need, and that all they required was a means of making them visible . . . Miracles were the norm for the people who walked so intimately with God."[96]

When Europeans began living in Indian villages in the seventeenth century, they made notes on American Indian beliefs about birds. French captain Pierre-Charles de Liette lived with the Peoria and Kaskaskia tribes on the Illinois River from 1687 to 1702. He described how kestrels, crows, vultures, mourning doves, ducks, swallows, martins, parrots "and many others" served as manitous, or spirit guides, for his hosts. According to de Liette, every Illiniwek warrior kept a sacred bird skin rolled in a specially decorated bundle and invoked its help before setting out on a dangerous mission:

> After eating with great appetite they all go to get their mats and spread out their birds on a skin stretched in the middle of the cabin and with the *chichicoyas* [gourd rattles] they sing a whole night, saying: "kestrel, or crow, I pray to you that when I pursue the enemy I may go with the same speed in running as you do in flying, in order that I may be admired by my comrades and feared by our enemies."[97]

The bird skin bags that Captain de Liette described three hundred years ago were carefully fabricated after their owners had gone on successful vision quests and acquired a spirit guide. "See to it," one Ho-Chunk father told his son as he set out on a vision quest, "that, of all those spirits Earthmaker created, at least one has pity upon you and blesses you. Whatever such a spirit says to you, that will unquestionably happen."[98]

Reverend Mails wrote that among Plains tribes, men and women alike were expected to seek at least one vision, "both for the power they received personally and for the cumulative power which came to the tribe through the totality of all the visions." Vision seekers typically went into seclusion in a wild location at or before puberty and repeated the experience as often as necessary in later life. Isolated in the wild far from other people, they washed and painted themselves, fasted for several days, concentrated their minds through prayer and walking meditation, and asked the spirits to help them. Though many seekers received no vision, most had faith that if their suffering was genuine and their prayers were focused and sincere, a vision would eventually come through something like divine grace.[99]

Birds were often encountered during these transcendental experiences. Here are three examples from Midwestern vision quests:

> *Lakota, South Dakota, ca. 1910:* "All at once I was way up there with the birds. The hill with the vision pit was way above everything. I could look down even on the stars, and the moon was close to my left side

. . . A voice said, '. . . We are the fowl people, the winged ones, the eagles and the owls. We are a nation and you shall be our brother . . . You will learn about herbs and roots, and you will heal people. You will ask for nothing in return. A man's life is short. Make yours a worthy one.'"[100]

Ho-Chunk, Wisconsin, nineteenth century: "Then I looked above and I saw that it was very cloudy, yet straight above me in a direct line the sky was blue . . . The Thunderbirds were blessing me. With the blue sky, they were blessing me . . . The Thunderbirds were the spirits speaking to me. They had spears and little war-clubs in their hands and (wreaths) made of flat cedar leaves upon their heads [and they gave the narrator the power to take and restore life]."[101]

Ojibwe, early twentieth century: "Soon he uncovered his face and looked out and saw a flock of geese that now sounded like men speaking. The boy had not had anything to eat nor drink for these eight days. The head one—there is always a head goose—spoke to the boy telling him never to have fear in any war, and should he be in danger in any war to think of this flock of geese and he would always come out unhurt."[102]

Conversations with mystical beings, out-of-body experiences, visual and auditory hallucinations, and a life-changing sense of deep authenticity are hallmarks of these shamanic visions. After returning home and describing the message given by a spirit guide, the vision seeker goes "forth with weapons or traps, until he can procure the animal or bird [he has seen], the skin of which he preserves entire, and ornaments it according to his own fancy, and carries it with him through life."[103]

A Pawnee bundle always contained at least a pipe, tobacco, paints, birds, and corn, as well as any other objects the owner thought would convey supernatural power. Birds of all kinds were also commonly part of Blackfoot bundles. An Arikara medicine bundle from Nebraska contained the skins of six now-extinct Carolina parakeets. Others contained various woodpeckers and owl species; one included an owl cleverly preserved in a perched position with eyes of polished buffalo horn. Some tribes held that

Hidatsa warrior offering incense to his medicine bundle, ca. 1908. Photograph by Edward S. Curtis. CHARLES DEERING MCCORMICK LIBRARY OF SPECIAL COLLECTIONS, NORTHWESTERN UNIVERSITY LIBRARIES

the spirit guides actually dwelled inside the artifacts in the sacred bundles, rather like communion wafers are said to contain the actual body of Christ or a Hindu statue the incarnation of Vishnu. Sacred bundles were opened with great care and proper ceremony, and were often buried with their owner at the end of his or her life.[104]

Some bundles belonged not to individuals but to clans, whose authority they legitimized. These were believed to have been given to humans centuries ago by spirit guides in the same way that the Ho-Chunk were given weapons by Thunderbirds at the end of the Red Horn narrative. Their contents often included feathers, talons, or even the entire head of a bird, as well as herbs, bones, and other spiritually charged objects. Unlike personal bundles, clan bundles were carefully passed down through successive generations.[105]

"Besides his medicine bundles, each warrior carried or wore a smaller and more convenient personal medicine—to remind him that his helper was always close at hand," wrote Reverend Mails. "Sometimes it was simply the skin of a helper bird which was tucked behind the ear before going into battles. Other times it was a bird made of rawhide . . . with a tie string for fastening it to the hair." Oglala Sioux leader Crazy Horse wore the full body of a hawk on his left shoulder when he rode into battle: "He was as sure and swift as the hawk himself . . . ," recalled Lakota chief Standing Bear, "and as fearless."[106]

Even a single feather from a bird could convey supernatural power to its owner. When Shawnee warriors in Ohio were threatened in 1778, they immediately "shaved their heads, ready to lay the plume on and turn out to war . . . ," reported missionary John Heckewelder. "That tuft of hair on their heads, termed the scalp, being daubed over with tallow, the white plume from the head of the eagle is stuck on; they say that this confers the courage of that bird on them."[107]

Owl feathers were worn in the 1870s by certain Lakota warriors who each believed that by "thus showing his respect for nature, he will be favored by nature by having his own powers of sight increased. It is a fact that the vision of the members of the Owl Lodge was exceedingly keen." This elite group consisted of warriors who had been especially brave or scouts who had risked their lives to retrieve information from behind enemy lines.[108]

Jonathan Carver, the first Englishman to reach the northern Plains, wrote in 1766 that "those among the men who wish to appear gayer than the rest, pluck from their heads all the hair except from a spot on the top of it about the size of a crownpiece, where it is permitted to grow to a considerable length: on this are fastened plumes of feathers of various colours with silver or ivory quills. The manner of cutting and ornamenting this part of the head distinguishes different nations from each other."[109]

No birds' feathers were so revered as those of eagles. Lakota chief Standing Bear wrote, "The eagle symbolizes to the Indian the greatest power . . . He flew so high in the sky that he reached the realm of the gods and could look the world over. So the Indian wore the feathers of the eagle long before the white man came to this land. And he wore them, not to 'look nice,' but with awe and appreciation of the wonder of nature." Early in the nineteenth century, Mandan warriors on the lower Missouri River would trade a valuable horse for the skin of a single golden eagle.[110]

Eagles' feathers were thought to carry sacred power more effectively than any other bird's. Golden eagles were considered more powerful than bald eagles because they were larger and stronger; their wingspan can stretch up to seven feet and they can reach speeds of 150 miles per hour when diving. Eagles were also believed to be closely related to the Thunderbirds, who pierced the sky with bolts of lightning from their eyes and shook the earth with thunder from their wings. Virtually all American Indian nations in the continent's interior held eagles in awe and incorporated them into stories, art, ceremonies, and daily life, and eagle feathers are still among the most revered by practitioners of Native religions.[111]

In order to obtain eagle feathers, Plains tribal members caught the eagles by hand. "The hunter digs a ditch on a broad, flat prairie," wrote fur trader Pierre-Antoine Tabeau in 1803,

> wide and deep enough for him to lie at his ease. He covers it with straw or with interlaced branches in such a way that, invisible himself, he can see the bird when it pounces upon a rabbit or other bait fixed on the surface. The bird, seizing the prey, finds itself caught by the feet and dragged to the bottom of the ditch. The same bait sometimes deceives many of them in a day. But it also often happens that

Hidatsa eagle trapper, ca. 1908. Photograph by Edward S. Curtis. CHARLES DEERING MCCORMICK LIBRARY OF SPECIAL COLLECTIONS, NORTHWESTERN UNIVERSITY LIBRARIES

> the hunter passes many days in the ditch without seeing a single bird and, if his industry is wonderful, his patience is no less so.[112]

If one or two eagles were captured, hunters usually released them after removing a few tail feathers. But if a large number came to the bait, some might be carefully strangled by hand. Their wings would be removed and brought back to the village while the rest of the body was left on a carefully prepared altar. Young eagles were even sometimes kept caged or tethered by the Mandan and Hidatsa so their plumes could be harvested.[113]

Eagle feathers were also worn not just to channel sacred energy, but also as marks of honor. "There is a language of feathers for the Indian," explained Chief Standing Bear.

> The way and the number of eagle quills that a warrior wears on his head have a meaning: The bravest warrior, who has gone ahead of the rest and led the way or faced the enemy and either touched him or killed him, wears a feather straight up at the back of his head. The next bravest man wears his feather pointing sideways to the right from the back of his head. The third bravest man wears his feather in the same way as the second bravest, only it points to the left side. The fourth bravest man wears his feather hanging straight down his back. A red stripe straight across the quill meant that the man wearing it had been wounded once by the enemy. Sometimes a warrior who had been on many war parties wore two and three or even four stripes on his feather. A brave man who had been surrounded by the enemy and had fought his way out wore a very pretty decoration on the side of his head. It was made of a tuft of feathers, from the center of which there hung a buckskin string, to the end of which there was tied a single feather. At large social gatherings and council meetings the warriors all wore their various decorations, and every one knew what deeds of bravery each man had performed by looking at the decorations.[114]

Traveling in Minnesota in June 1823, Italian count Giacomo Beltrami noted that during ceremonial dances, Ojibwe warriors

A group of Sauk and Fox (Mesquakie) Indians, ca. 1833, showing their use of feathers to mark honor and status. Hand-colored engraving by Karl Bodmer. WHI IMAGE ID 4519

> all carry their bow, quiver, and arrows; as well as a plume of feathers on their head, the exclusive distinction of warriors of renown. The feathers are from a bird which the Canadians call *killiou*, and the Indians *wamend-hi* [golden eagle]. These birds are so rare and so highly valued by the Indians in general, that whoever has the good fortune to kill one of them, receives the formal compliments of the whole camp on his success, and is entitled to the privilege of wearing one of its plumes. Every warrior is authorised to wear as many of them as he has killed of his enemies; and every time he destroys one of these birds, he adds a plume to his previous honours.[115]

Eagle feathers were also widely used in ceremonies. The annual spring ritual performed by the Mandan to call buffalo herds included dancers attired in eagle quill skirts and crowns. At his infant naming ceremony, Ojibwe elder Kahkewaquonaby was given "a bunch of eagle feathers prepared for the occasion. It was considered sacred, as it represented the speed of the thunder and the eagle." Countless other examples could be cited.[116]

"Feathers were used in many ways, both in their natural colors and dyed," recalled Standing Bear of the Sioux in the 1870s, "but the beautiful

Amiskquew, a Menominee warrior, in 1832. Hand-colored lithograph by Charles Bird King for McKenney and Hall's *Indian Tribes of North America*. WHI IMAGE ID 27677

Na-she-mung-ga, a Miami chief, in 1825. Hand-colored lithograph by James Otto Lewis. WHI IMAGE ID 26876

eagle-feather headdress was the crowning piece of Lakota regalia, its splendid appearance adding to the attraction of even the handsomest warrior." Sometimes called a "warbonnet," this majestic trailing headdress became an icon in mainstream American media, symbolizing to white audiences not just Plains Indians but all American Indians, despite the fact that most Indian nations did not wear it.[117]

One of the earliest images of a Plains eagle-feather headdress shows Mandan chief Mah-to-toh-pa, or Four Bears, in South Dakota in 1832. "His breast has been bared and scarred in defence of his country," wrote artist George Catlin, who interviewed and painted the chief, "and his brows crowned with honours that elevate him conspicuous above all of his nation. There is no man amongst the Mandans so generally loved."[118]

"The Head-dress," Catlin continued, "which was superb and truly magnificent, consisted of a crest of war-eagle's quills, gracefully falling back from the forehead over the back part of the head, and extending quite down to his feet; set the whole way in a profusion of ermine, and surmounted on the top of his head, with the horns of the buffalo, shaved thin and highly polished." Each Plains nation had its own preferred way to make and wear the large feathered headdress.[119]

The many ways in which dozens of American Indian peoples have worn feathers on their heads, and what they signified, could fill an entire vol-

Mandan chief Mah-to-toh-pa in 1832. Hand-colored lithograph by George Catlin. WHI IMAGE ID 69159

ume. Some headgear was made from whole skins of birds, or included entire heads of birds. Some tribes attached clusters of feathers to fur or leather caps. Some wore a simple crown of erect plumes, while others wore a single feather all by itself. A wide variety of combinations, colors, numbers, and arrangements of feathers can be found among the early-nineteenth-century portraits of Indians painted by James Otto Lewis, Thomas McKenney and James Hall, George Catlin, Karl Bodmer, and other artists who traveled the frontier, and each arrangement communicated the kinds of specific information that Standing Bear outlined above.[120]

In addition to headdresses, Native peoples wore feathers and bird skins in other ways, particularly on special occasions. "The Lakotas were very proud of feather-trimmed garments and ornaments," recalled Chief Standing Bear, and they maintained them carefully. "Crisp cold air," he explained, "enlivens both furs and feathers and a bedraggled breath plume will come to life in a short time under the magic touch of air." Elite Sioux women wore the soft downy feathers of eagles in their hair; these were also attached to wands used in their corn dance. In the 1660s, the Shawnee in the Ohio Valley wove scarves and belts from the brilliant green and yellow feathers of the Carolina parakeet. Ojibwe men in northern Minnesota made light waterproof hats from the pouches of pelicans. Hundreds of examples of Indian uses of birds for dress or decoration could be cited.[121]

Besides using feathers for ceremonies and adornment, American Indians around the Midwest also used them in more utilitarian ways. Ohio missionary John Heckewelder wrote in about 1780 that

> the blankets made from feathers were also warm and durable. They were the work of the women, particularly of the old, who delight in such work, and indeed, in any work which shows that they are able to do their parts and be useful to society. It requires great patience, being the most tedious kind of work they perform, yet they do it in a most ingenious manner. The feathers, generally those of the turkey and goose, are so curiously arranged and interwoven together with thread or twine, which they prepare from the rind or bark of the wild hemp and nettle, that ingenuity and skill cannot be denied them.

Mesquakie chief Wah-Pel-La, wearing a feathered shawl, ca. 1836. Hand-colored lithograph by Charles Bird King for McKenney and Hall's *Indian Tribes of North America*. WHI IMAGE ID 28365

Many other Native peoples, including the Iroquois and Mesquakie (or Fox), also wove feathers into blankets, robes, and cloaks.[122]

Birds were also made into tools and utensils. In the 1790s, Moravian missionary George Henry Loskiel noted that the Shawnee in the Ohio Valley made pouches from waterproof loon skins "large enough to hold their pipe, tobacco, flint and knife, &c." On the Great Lakes, loon skins were used "to cover their guns, to prevent the wet from spoiling them." In the 1670s, Huron girls informed Jesuit missionary Claude Allouez that they slept with "Eagles Talons hanging at their sides as a soldier carries his sword, wherewith to defend themselves (they told me) against the insolence of the young Men." Whistles and flutes fashioned from the long wing or leg bones of large birds, especially eagles, were used to communicate on the battlefield or call warriors into action. Bone tubes were also widely used by shamans to suck malicious spirits out of sick patients, and goose grease was rubbed onto sore muscles and joints as a liniment.[123]

All across the region, fans were made from wings or feathers of turkeys, eagles, cranes, and other large birds to clear campfire smoke from inside lodges and tipis or to ritually cleanse participants in ceremonies. A white prisoner who spent six years with the Shawnee in Ohio reported to James Kenney:

> When they Shoot one of [the eagles], that they immediately cut out his tongue & his heart & hide them in some rotten Log & hangs up ye Body for two Days, then bring it home, use ye wings, (took carefully off & join'd with some piece of white Linnen at ye Roots) to raise wind, as he says, which he has seen done, goes in some private place & fands with ye wings Singing some song all ye time for ye Space of near two Hours or more, when ye wind Will Raise ready to blow down ye Houses.[124]

No bird-associated implement was so important to Indian nations throughout the interior of the continent as the calumet, the ceremonial pipe used to guarantee peace between strangers or adversaries. On June 25, 1673, while descending the Mississippi, Father Jacques Marquette and fur trader Louis Joliet visited a village of Peoria Indians near the mouth of the Des Moines River in northeast Missouri. They were greeted by elders bearing a calumet. "There is nothing more mysterious or more respected among them," Marquette wrote.

> Less honor is paid to the Crowns and scepters of Kings than the Savages bestow upon this . . . They also use it to put an end to Their disputes, to strengthen Their alliances, and to speak to Strangers . . . It is fashioned from a red stone, polished like marble, and bored in such a manner that one end serves as a receptacle for the tobacco, while the other fits into the stem; this is a stick two feet long, as thick as an ordinary cane, and bored through the middle. It is ornamented with the heads and necks of various birds, whose plumage is very beautiful.

> To these they also add large feathers—red, green, and other colors—wherewith the whole is adorned.

In September 1679, explorer Louis Hennepin described his experience being greeted with a calumet four hundred miles to the northeast, by the Potawatomi at Green Bay, Wisconsin:

> They tie to it two Wings of the most curious Birds they find, which makes their Calumet not much unlike Mercury's Wand, or that Staff Ambassadors did formerly carry when they went to treat of Peace. They sheath that Reed [the pipe-stem] into the Neck of Birds they call Huars [loons], which are as big as our Geese, and spotted with Black and White; or else of a sort of Ducks who make their Nests upon Trees, though Water be their ordinary Element, and whose Feathers are of many different Colours [wood ducks]. However, every Nation adorns the Calumet as they think fit according to their own Genius and the Birds they have in their Country.

The calumet was used to solemnize treaties by nations living in the north from the Great Lakes to the Rocky Mountains, and south to Mississippi and Arkansas. "It seems to be the God of peace and of war, the Arbiter of life and of death," wrote Marquette in 1673. "It has but to be carried upon one's person, and displayed, to enable one to walk safely through the midst of Enemies—who, in the hottest of the Fight, lay down Their arms when it is shown. For That reason, the Illinois gave me one, to serve as a safeguard among all the Nations through whom I had to pass during my voyage."[125]

A century and a half later, George Catlin found the tradition of the calumet still firmly observed all across the Midwest and Plains:

> The calumet, or pipe of peace, ornamented with the war-eagle's quills, is a sacred pipe, and never allowed to be used on any other occasion than that of peace-making: when the chief brings it into treaty, and unfolding the many bandages which are carefully kept around it—has it ready to be mutually smoked by the chiefs, after the terms of the treaty are agreed upon . . . The mode of solemnizing is by passing the sacred stem to each chief, who draws one breath of smoke

Ho-Chunk chief Naw-Kaw. Hand-colored lithograph by Charles Bird King for McKenney and Hall's *Indian Tribes of North America*. WHI IMAGE ID 2287

> only through it, thereby passing the most inviolable pledge that they can possibly give, for the keeping of the peace. This sacred pipe is then carefully folded up, and stowed away in the chief's lodge, until a similar occasion calls it out to be used in a similar manner.[126]

Other pipes used simply for social and recreational smoking might also be decorated with feathers or bird motifs. In the illustration by Catlin on the opposite page, the calumet with its suspended fan of large feathers is shown at the top and nonceremonial pipes below it.

Of course, feathers could be used for the preceding purposes only if birds were hunted and captured. Before adopting the bow and arrow about

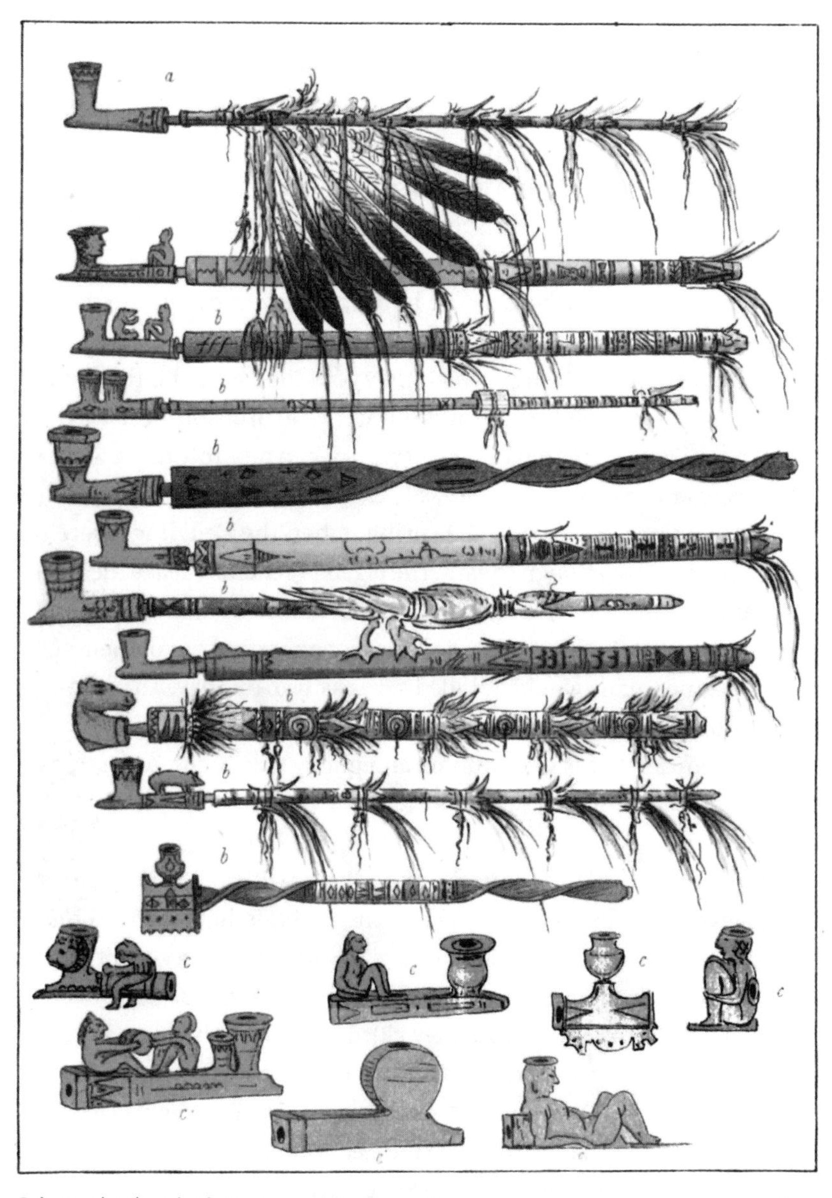

Calumet (top) and other pipes used in the 1830s. FROM *NORTH AMERICAN INDIANS, VOL. 1* BY GEORGE CATLIN

500 CE, Native peoples in the Midwest caught birds in their bare hands, clubbed them with sticks, and snared them in nets. It's likely that they also took down larger species such as turkey, swans, and cranes using atlatls and darts. These techniques required a hunter to move calmly and quietly close to his prey, much closer than a modern birder with binoculars usually does.

This was easier with some species than others. Ruffed grouse were so unsuspecting that a seventeenth-century Canadian priest reported, "One can drive them like chickens before him; and they even allow themselves to be approached near enough to permit one to extend a noose attached to the end of a pole, through which they pass the head, and thus render easy this method of capture." Native hunters had probably been catching them in similar ways for ten thousand years when he wrote those words.[127]

Passenger pigeons, too, were easily caught by hand. Enormous numbers were vulnerable during the spring, when the fledglings were too young to fly but big enough to eat. "The pigeons generally make their nests near human habitations," a Ho-Chunk man recalled early in the twentieth century. "Long poles were taken and the pigeons poked out of their nests. In this manner many would be killed very easily in one day. They are then either broiled or steeped, when they have a delicious taste."[128]

Sometimes pigeons were hunted at night by torchlight. A missionary in the Ohio Valley in the 1790s reported:

> At night a party of Indians frequently sally out with torches made of straw or wood, and when they get among the birds, light them. The pigeons being dazzled by the sudden glare, are easily knocked off the branches with sticks. Such a party once brought home above eighteen hundred of these birds, which they killed in one night in this manner. Their flesh has a good taste, and is eaten by the Indians either fresh, smoked, or dried. When the Iroquois perceive that the young pigeons are nearly fledged, they cut down the trees with the nests, and sometimes get two hundred young from one tree.[129]

Dead pigeons were often preserved for later use. In 1794, a missionary in western New York and Pennsylvania reported that the Iroquois smoked and dried pigeons for later consumption, as did the Ho-Chunk on the Wisconsin shore of Lake Michigan in 1820. Traveling up the Mississippi in the spring of 1806, Zebulon Pike encountered four canoes of Sauk Indians "with wicker Baskets, filled with the young pigeons."[130]

Capturing other birds by hand usually required great patience and often involved decoys. Baron Louis-Armand de Lahontan came to the western Great Lakes in 1687, where he saw Indians use blinds to catch waterfowl:

> These Water-Hutts are made of the branches and leaves of Trees, and contain three or four Men: For a Decoy they have the skins of Geese, Bustards [Canada geese], and Ducks, dry'd and stuff'd with Hay, the two feet being made fast with two Nails to a small piece of a light plank, which floats round the Hutt. This place being frequented by wonderful numbers of Geese, Ducks, Bustards, Teals, and an infinity of other Fowl unknown to the Europeans, when these Fowls see the stuff'd Skins swimming with the Heads erected, as if they were alive, they repair to the same place, and so give the Savages an opportunity of shooting 'em, either flying, or upon the Water; after which the Savages get into their Canows and gather 'em up.[131]

An early-nineteenth-century prisoner reported that his American Indian captors lured turkeys close by "having prepared from the skin an apt resemblance of the living bird . . . This management generally succeeds to draw off first one and then another from their companions, which, from their social and unsuspecting habits, thus successively place themselves literally in the hands of the hunters."[132]

Chief Simon Pokagon, a Potawatomi elder raised in Wisconsin and Michigan in the 1820s, recalled an Indian hunter capturing waterfowl with camouflage mounted on his head:

> As we were packing our goods into his boat, I noticed a basket of curious make, with wild rice trimmed, as if growing in the water . . . He put the basket over his head, and jumped into the stream, the water just covering his shoulders. Slowly he walked in the water toward the ducks; the box trimmed with rice straw appeared to be floating toward the flock, soon moving among them. They appeared to realize no fear of danger from the floating craft [and soon] I saw one duck

> sink like lead out of sight; then another disappeared, leaving scarcely a sparkle where they sank.[133]

His companion had simply seized them by the feet and strangled them under the water.

Nets were used by Native hunters to trap large numbers of birds all at once. Fur trader Nicolas Perrot described how tribes in the forests around the western Great Lakes captured passenger pigeons in the late 1680s:

> They make in the woods wide paths, in which they spread large nets, in the shape of a bag, wide open, and attached at each side to the trees; and they make a little hut of branches, in which they hide. When the pigeons in their flight get within this open space, the savages pull a small cord which is drawn through the edge of the net, and thus capture sometimes five or six hundred birds in one morning, especially in windy weather.[134]

Nets were also weighted and sunk underwater to capture ducks and geese that landed on wild rice beds. Seventeenth-century Jesuit missionaries described how, at modern Green Bay, Wisconsin, "The Savages stretch nets in certain places where these Fowl alight to feed upon the wild rice. Then advancing silently in their Canoes, they draw them up alongside of the nets in which the birds have been caught."[135] Another priest there described how,

> without counting the fish, they sometimes catch in one night as many as a hundred wild fowl. This fishing is equally pleasant and profitable; for it is a pleasure to see in a net, when it is drawn out of the water, a Duck caught side by side with a pike, and Carp entangled in the same meshes with Teal. The Savages subsist on this manna nearly three months.[136]

Despite harvesting some species by hand or in nets, most Native peoples shot birds with the bow and arrow. While camped in Minnesota in September of 1699, Pierre-Charles Le Sueur noted that the Sioux "handle the bow very expertly, and were several times seen to shoot ducks on the wing." His contemporary Pierre Radisson saw an Ojibwe hunter kill three ducks with a single arrow.[137]

Use of the bow was taught to children "as soon as they are able to run about," noted an eighteenth-century observer in Ohio. "When they grow older they shoot pigeons, squirrels, birds and even raccoon with their bow and arrows." English trader John Long noted in about 1770 that Ojibwe youths could "kill small birds at fifty yards' distance, and strike a half-penny off a stick at fifteen yards." Potawatomi Chief Pokagon claimed to have watched a boy and a girl simultaneously shoot arrows through the neck of the same distant swan. Almost any bird that was valued for culinary, ceremonial, or decorative purposes could be obtained with bow and arrow.[138]

Native hunters mastered guns within a generation of encountering them. The Iroquois in New York and the Huron in the eastern Great Lakes possessed firearms by the 1640s and tribes in modern Michigan, Wisconsin, and Illinois by 1670. Those west of the Mississippi had firearms by the end of the seventeenth century, and nations on the eastern Plains possessed them by the mid-1700s. A Dutch observer noted in 1650 that Iroquois hunters were "so skilful [*sic*] in the use of them that they surpass many Christians." Countless generations of hunting with the bow and arrow had equipped Native peoples with powers of tracking, concentrating, aiming, and shooting that amazed white settlers.[139]

For example, an Englishman traveling in Ohio in the 1770s wrote:

> I remember seeing some Americans shooting at a loon, a bird nearly the size of an English goose. This bird is remarkable for diving, and generally rises some yards from the place where it dips. They fired at the distance of one hundred and fifty yards with a rifle, several times without success: an Indian standing by, laughed at them, and told them they were old women: they desired him to try his skill, which he instantly did: taking his gun, and resting it against a tree, he fired, and shot the loon through the neck.[140]

On a fall day in about 1850, a Wisconsin logging foreman named Sam Barker challenged an Ojibwe hunter to hit a particularly high flock of approaching geese. The hunter, a chief named Opwaganes, was born in 1788 at Lac Court Oreilles near Lake Superior and was reputed to be the best shot in the North Woods. "I want that big gander on the lead of that flock for my dinner," Barker told him. "I want the top of his head snipped off; don't shoot him through the body and spoil the meat or I won't have it." Winking at the crowd of lumberjacks that had assembled, Barker added, "Here is a $10 greenback for you if you do a good job." In 1850, $10 was worth about $250 in today's money. Opwaganes lifted his gun, aimed, and fired, and the dead goose was soon retrieved, minus the top of its head. The Ojibwe chief silently tucked away the cash and departed into the forest.[141]

Although their technology changed during these decades, American Indian hunters continued to employ traditional strategies. John James Audubon accompanied a hunting party in Ohio in December 1810 that split in half to hide on opposite shores of a lake covered in migrating swans. "Being divided," he explained, "three on one side and four on the other, the former hid themselves and when the birds flew from the latter, they alighted within good distance of those who had first alarmed them." Taking turns as the swans alternately fled from one side of the lake to the other, the Indians quickly shot fifty of them in order to sell the skins to white traders.[142]

Regardless of how they were captured, most birds immediately became a meal. Indians typically broiled large species such as turkeys, swans, or geese by hanging their fleshy parts over an open flame. Smaller birds such as ducks or grouse were impaled whole on spits, which were then stuck into the earth before a fire. Indian cooks also routinely added birds to boiling corn or wild rice to make a stew. "Of the wild ducks and Indian corn," recalled an English prisoner in Ohio in 1764, "we made broth; the Indians made spoons of the bark of a tree in a few minutes, and, for the first time, I eat of boiled wild duck. When we marched on after dinner, I could perceive no fragments left."[143]

Another prisoner of Kickapoo warriors in Ohio in the spring of 1788 recalled:

> [The Indians] killed two ducks that evening, the ducks were very fat, they picked one of the ducks, and took out all its entrails very nice and clean—then stuck it on a stick, and stuck the other end of the

> stick in the ground before the fire, and roasted it very nice. By the time the duck was cooked, one of the Indians went and cut a large block out of a tree to lay the duck upon; they made a little hole in the ground to catch the fat of the duck while roasting; when the duck was cooked they laid it on this clean block of wood, then took a spoon and tin cup, and lifted the grease of the duck out of the hole and took it to the cooked duck on the table, and gave me some salt, then told me to go and eat . . . I thought I had never eat any thing before that tasted so good. That was the first meal I had eaten for four days. The other duck they pulled a few of the largest feathers out of, then threw the duck, guts, feathers, and all, into their soup-kettle, and cooked it in that manner.[144]

Indians also roasted entire birds. When visiting an elder near Lake Michigan in about 1830, Potawatomi chief Simon Pokagon was treated to a first course of pigeons broiled on spits, and, "When I had supposed the meal was eaten, the old man raked from the embers several birds wrapped in large, wet leaves, undressed and unpicked, saying, 'Me-no o-me-me-og (Good pigeons) cook all day.'" Stripping off the feathers and skin, Pokagon "was fully satisfied that I had never before eaten meat more pleasing to the taste." Looking back on his childhood in the 1860s, Shawnee elder Thomas Alford reflected that "a duck or squirrel wrapped in wet corn husks and baked or roasted in hot ashes, cannot be surpassed, according to my notion."[145]

Beyond their utilitarian value, hunting, eating, and wearing birds were all activities suffused with religious connotations for Native peoples. "All birds were said to have some spiritual power," Reverend Mails wrote of the Plains nations.

> The eagle, hawk, and owl—all birds that captured their prey—typified the courage and dash needed for success in war . . . The raven, magpie, and chickadee were birds of great wisdom. They also talked to people, telling them of coming events, leading them to game, advising them of danger, and recommending a course of action. Certain small water birds were used as messengers by the supernatural powers.[146]

It wasn't unusual for indigenous people to believe that prairie chickens, meadowlarks, crows, and other species spoke to humans. The nineteenth-century Lakota chief Standing Bear wrote that his people "knew, of course, that the animals and birds had a way of talking to one another just as we did. We knew, too, that the animals and birds came and talked to our medicine men . . . The larks in our state, at that time [ca. 1860] talked the Sioux language—at least we inferred that they did, but in California, where I now live, it is impossible to understand them."[147]

Perhaps most importantly, birds brought crucial information from the spirit world. Jonathan Carver wrote in 1766, of the whip-poor-will: "As soon as the Indians are informed by its notes of its return, they conclude that the frost is entirely gone, in which they are seldom deceived; and on receiving this assurance of milder weather, begin to sow their corn." But tribes from New York to Minnesota also thought the whip-poor-will was "a bird of evil omen, and that if [it] light on a house, the death of some of the inhabitants is inevitable."[148]

Indians in the Midwest saw birds this way because they embraced the basic tenets of their shamanic heritage as common sense: that what we learn through our senses does not encompass all knowledge, that the world is alive with supernatural forces, and that each person must live a proper life to help keep the universe in balance. Seeing Canada geese migrate overhead, hearing the rumble of distant thunder, watching a whip-poor-will flit above a lodge, or listening to meadowlarks warble in the prairie grass confirmed Native peoples' place in the universe.

This was, of course, a very different perspective than that carried into the interior of the continent by white explorers and settlers. In May 1791, American colonel Thomas Proctor complained that Seneca leaders refused to meet with him to discuss diplomatic affairs because "it was their pigeon time, in which the Great Spirit had blessed them with an abundance; and that such was his goodness to the Indians that he never failed sending them season after season, and although it might seem a small matter to me, the Indians will never lose sight of those blessings."[149]

To the Seneca, the pigeons were not just a commodity, but brethren—fellow creatures inhabiting a world whose welfare depended on maintaining cosmic balance through proper ceremonies and behavior. The birds arrived once a year in late spring just as food stored over the winter was running out and before new crops had matured. To the Seneca, gathering, eating, and giving thanks for the pigeons was obviously much more important than white people's politics.[150]

Belief in the sacredness and spiritual power of birds did not disappear from Native communities when science became orthodox in mainstream America. Today, "countless Native Americans have these beliefs and see birds as messengers, or oracles, to use a European term," writes Micmac author Evan Pritchard. He interviewed American Indian elders and shamans around the nation for his 2013 book, *Bird Medicine*, concluding that "it is clear that a living, dynamic tradition still exists concerning our spiritual relationship with birds."[151]

Pritchard met holy men and lay practitioners who still use traditional shamanic methods to

> drop the barrier that has grown up between animals and humans in the last five thousand years. When it drops, the medicine person sees what birds see, hears their thoughts, feels as they feel, and understands their concerns. In this open state, according to the shaman's solitary path, they see the past and the future and see what is going on at a distance and in the spirit world . . . They can communicate with the overlighting spirit of an animal or bird and ask for its help.[152]

I had some insight into such things after my second child was born. Mary and I tried to help each other preserve a little private time amid the chaos of parenting two little kids, and one Saturday afternoon she pushed me out the door to my favorite wetland. The babies had been screaming, she and I had quarreled, and as I walked along the trails I found no birds, adding disappointment to an already frustrating day. Everything was wrong. I knelt down in the grass, closed my eyes, and just gave up.

Five minutes later, I opened my eyes to find birds all around me—tree swallows above the meadow grass, redwings in the cattails, yellow warblers in the brush, gulls overhead. They'd been there all along, but my awareness had been focused exclusively on the negative voices in my head rather than on the world around me. Trivial as it is, the experience suggests that disciplined practitioners who can drop the barriers of rational thought and language may connect with nature very differently than I do. As the spiritual teacher Ram Dass likes to say, "The quieter you become, the more you can hear."

One of Pritchard's elderly informants summarized the Native tradition by telling him, "Birds are close to the Indian heart. I love them because they have such wisdom. Why are they wise? The highest flying bird is closest to the Creator's ways. When they are close to you, something happens to make things better. Either they bring you goodness, or they warn you. If you truly believe they can help you, the bird knows what you are thinking and can help you in many ways."[153]

My father died the spring following my little epiphany in the marsh. After multiple cross-country trips to sit beside his comatose body, negotiate with doctors, support my family, arrange a cremation, write his obituary, and deal with my own grief, his funeral finally brought some closure.

A few days afterward, Mary and I went out birding for the first time in months. As we walked the dirt road beside an upland meadow, a solitary bluebird followed us, flitting from bush to telephone wire to tree branch to shrub. This was nearly thirty years ago and bluebirds were uncommon in our part of the Midwest. I'd never seen one at this place before, and though it seemed crazy, as I watched it, I felt a burden lift, as if the bluebird were a sign that my father was all right.

Several weeks later we were a couple of miles away on another trail when a flock of bluebirds, parents and offspring, whirled out of the sky and settled in a shagbark hickory nearby. They stayed only a moment before taking wing again, the mixed flock of parents and offspring reminding me that not only was my father okay, but that life always comes full circle.

These realizations were not ideas or beliefs, but feelings hardly expressible in words. Were they merely coincidences? Was I projecting my inner turmoil onto objective nature? Would a shaman see something else? Not being a shaman, I remain agnostic, as I did with the mushrooms in Harvard Yard long ago. Maybe real, maybe not.

5

Missionaries, Monsters, and Miracles

As Christopher Columbus sailed into the Caribbean in October 1492, he knew that land was near because for three days and nights his ships passed through a great wave "of birds flying from N. to S.W. . . . There were many land-birds, and they took one that was flying to the S.W." English captains were soon coasting the North American shoreline from Virginia to Labrador, and by 1534 French explorer Jacques Cartier had sailed up the St. Lawrence as far as Montreal.[154]

Europeans reached the heart of the country five years later, in 1539, when Spanish conquistadors stormed into Oklahoma and Kansas from bases in Florida and New Mexico. The French ascended the Great Lakes in the early seventeenth century and in 1622, before the Pilgrims at Plymouth had ventured beyond the scent of the Atlantic Ocean, they had circled Lake Superior. While the British huddled on the seashore, French soldiers, traders, and priests were coursing down the Ohio, exploring the Mississippi from Minnesota to Tennessee, and pushing west onto the Great Plains.[155]

Wherever they went, the first Europeans to enter the Midwest were greeted by tribal peoples who related to birds in the ways you've read about in the preceding chapters. For example, conquistador Hernan de Soto was met in 1539 by a chief wearing a "mantle of feathers down to his feet, very imposing," and another wearing on his head "a diadem of plumes." When he reached the Mississippi River, de Soto met "two hundred canoes filled with men, having weapons. They were painted with ochre, wearing great bunches of white and other plumes of many colours, having feathered shields in their hands." Farther west, a Plains Indian told Coronado that these Mississippi fleets were composed of "very big canoes, with more than 20 rowers on a side, and that they carried sails, and that their lords sat on the poop [stern] under awnings, and on the prow they had a great golden eagle."[156]

The newcomers assumed at first that North American birds resembled the birds back home. For example, one of de Soto's soldiers reported seeing "sparrows like those in Castille," and Jacques Cartier listed sixteen North American species that included "canaries, linnets, nightingales, sparrows and other birds, the same as in France."[157] But Europeans eventually realized that the "linnets" and "nightingales" were not, in fact, the same as those in Castile or Normandy. One of the first to make this discovery was French missionary Gabriel Sagard, who came to the Great Lakes in 1623.

Most of Sagard's life is shrouded in mystery. Even his birth and death dates are unknown, and nearly all we can say is that he was active between 1614 and 1636. He was a monk of the Recollect branch of the Franciscans, a religious order founded in 1588 and known for its vows of extreme poverty. After being assigned to missionary work in America in 1623, Sagard and another mendicant walked barefoot from Paris to Dieppe to board ship for the New World.[158]

It's impossible not to like Sagard. He combined an eager, childlike curiosity with extraordinary humility. Many Renaissance explorers' memoirs overflow with swagger and bombast, but Sagard's is a lesson in modesty. While watching seabirds on his voyage to the New World, he was puzzled about "where they can make their nests and hatch their young, being so far from land," but he concluded to "leave the matter to the decision of those who are wiser than [him]." His American Indian hosts loved him, and asked him not to leave at the end of his stay. His inquisitive, open-minded attitude permeates the book he wrote after returning to France, including his chapter on American birds.[159]

Sagard left Paris on March 18, 1623, and reached Quebec in mid-July. He stayed there two or three days before heading west into the wilderness to try to save Huron souls. He spent most of the next year traveling among their villages on Lake Huron, tagging along on hunting and fishing trips, ministering to the sick, preaching to anyone who would listen, and observing everything about this strange new world.

After returning to France in the fall of 1624 he wrote down his impressions in *Le Grand Voyage du Pays des Hurons*, a four-hundred-page memoir published in Paris in 1632. Four years later he incorporated it in a massive *Histoire du Canada et Voyages que les Frères Mineurs Recolects y Ont Faicts pour la Conversion des Infidelles*. The second part of Sagard's memoir opens with a long description of "the land and water animals and of the fruits, plants and natural abundance found ordinarily in the country." It includes eight pages on the birds of the eastern Great Lakes, the most thorough description of North American birds published up to that time.[160]

Sagard didn't try to catalogue every species the way scientists who came after him did. He simply wanted to show birds off as examples of God's glory. "It would be very difficult and unnecessary to describe all the species of birds which are in these extensive and vast Provinces," he wrote. "Those few which I have described will suffice to show that the Sky has its inhabitants for praising God just as we have our own here, & that everywhere are resounding praises for the Creator."[161]

Sagard started his bird accounts with

> the most beautiful, the rarest, and the smallest bird in the world perhaps, the Vicilin or hummingbird, which the Indians call in their language the Resuscitated. The body of this bird is no bigger than a cricket, the beak is long and slender, as thick as the point of a needle, and its legs and feet are as small as the lines in handwriting. Once the nest with the birds in it was weighed and found to weigh not more than twenty-four grains [about 1/20 of an ounce]. It feeds on dew and on the scent of flowers, not alighting upon them, but merely hovering over them. The plumage is as fine as down, and very pleasing and pretty to look at for the variety of its colouring.

He repeated a claim made by some Spanish writers that hummingbirds did not migrate but hibernated in trees: "This bird, they say, dies, or, to speak more accurately, goes to sleep in the month of October, remaining attached by its feet to some little tree-branch, and awakens in the month of April when flowers are plentiful."[162]

Sagard noted that many American birds were remarkably unafraid of humans:

> When the French first went into Canada, they found birds of all species so easy to take that anyone who had not seen it would not believe it; they would knock them from the trees with sticks, as I have seen the Indians do on the islands of Lake Huron, where we were camped for fish; and the partridges were so tame they let them put a loop around their necks on the end of a rod. When going after game, the hunter is sure to get as much as he can carry because they have never yet seen our muskets.[163]

One day, he said, a ruffed grouse "came right up to me in a corner where I was reciting my Office. And having looked me straight in the face went off slowly as it had come, spreading out its tail like a little peacock, and turning its head back continually; it looked at me and studied me quietly without any fear, and I indeed did not try to frighten it or put my hand upon it as I might have done, but allowed it to go off."[164]

Sagard was surprised that his Huron hosts didn't domesticate the wild turkeys that lived all around them. "If the savages were willing to give themselves the trouble of feeding young ones," he reflected, "they would domesticate them as well as we do here [in France], and likewise bustards or wild geese . . . but they make little effort to catch them, and so they enjoy them little & get even less nourishment from them . . . they have no domesticated animals other than dogs, and a few bears and eagles." The bear cubs were fattened for eating and the eagles were probably kept for their feathers, which were used for guiding arrows, in religious ceremonies, and as personal adornment.[165]

On a canoe trip back to Quebec, Sagard jotted down:

> There are also many eagles, which in their language they call Sondaqua. They usually make their nests on the banks of rivers or at the edge of some precipice, and quite at the top of the highest trees or rocks, so that it is very difficult to get at them and to discover the nests. Nevertheless we did discover several nests, but we found no more than one or two eaglets in any one. I thought of keeping some alive while on the way from the Hurons to Quebec, but both because they were too heavy to carry and also because I was unable to provide the fish they

> required (having nothing else to give them) we made a meal of them and found them very good, for they were still young and tender.[166]

As that anecdote suggests, missionaries endured long periods with only the simplest foods or none at all, and Sagard was often ready to eat anything, even crows. But crows and ravens played prominent roles in Huron religion, ceremony, and art, and the Indians refused to molest them. Sagard said:

> [I am] astonished that our Hurons do not eat the crow, which they call Oraquan, (which I would have no difficulty eating if I could catch one); there is nothing so nasty in this country, it gives them such horror. They don't even drive them from their grain, where they scratch like chickens, often making great trouble for us, and we lack the means to drive them off, which would be very difficult without continual warfare.[167]

Although Sagard clearly loved the New World, he necessarily saw it through European eyes. He usually expressed his praise or amazement in traditional Christian terms, as when he wrote that the habits of the goldfinch "made me like to watch them frequently and to thank God for their beauty and sweet warbling."[168]

Other times, his careful descriptions echoed the Greek or Roman texts he'd read as a student. For example, watching the aggressiveness of sandhill cranes, he was reminded of the *Iliad*:

> I am not at all surprised if it be true, what authors wrote about Cranes making war on Pygmies, who are small men about the height of a cubit living near the source of the Nile; because they are so large and strong that without a club a man would be perfectly unable to overcome them . . . Our French kill them with their muskets, more than the Natives with their arrows, but I assure you that they often find themselves effectively prevented from fighting; those which feel themselves struck go straight at the men in order to disfigure them, and they run as fast as a man.[169]

Sagard also told a story about eagles that used a motif he would have known from studying Greek classics:

> It once happened that one of our Priests, being alone in the woods about a league from Our Convent of Quebec, a very large Eagle or maybe a Griffon, swooped down on him with such fury that the poor Priest promptly holed up in a big bush, belly against the ground; the bird not being able to have its prey, batted its wings for a long time over the bush & then was compelled to give up, whereupon the Priest gave thanks to God.[170]

Sagard made no attempt to investigate, describe, or analyze American birds the way a modern ornithologist would. The fledgling scientific revolution had not penetrated his consciousness, and when he watched a loon plunge beneath a lake or an osprey soar overhead, he saw the handiwork of God. When he appealed for evidence to back any claim about nature, he was likely to cite the two-thousand-year-old works of a Greek or Roman author rather than the empirical evidence in front of his own eyes.

After twelve months in the wilderness, Sagard set out for Quebec to secure another year's supplies for his mission on Lake Huron. But when he had been in the city a few days, a letter arrived from his superiors at Paris ordering him to come home on the next ship. He wanted to return to the western Indians, "but God, praiseworthy in all things, without whose permission not a single leaf falls from a tree, willed that matters should turn out otherwise."[171]

Sagard's Huron guides made him promise to return in twelve months if he could, when they would again be downriver at Quebec, but they all knew it was unlikely. With a heavy heart, he sailed away in October 1624, arriving safely at Dieppe and returning to Paris as he had left it, barefoot and vowing to humbly serve God. He never returned to America, and almost nothing is known about the rest of his life.[172]

Four French missionaries had come to North America before Sagard, but three hundred would follow him over the next century and a half. Sagard and those first four were Recollects, but after 1632 the Jesuit order held a near-monopoly on proselytizing in Canada, the Great Lakes, and the Mississippi Valley.[173]

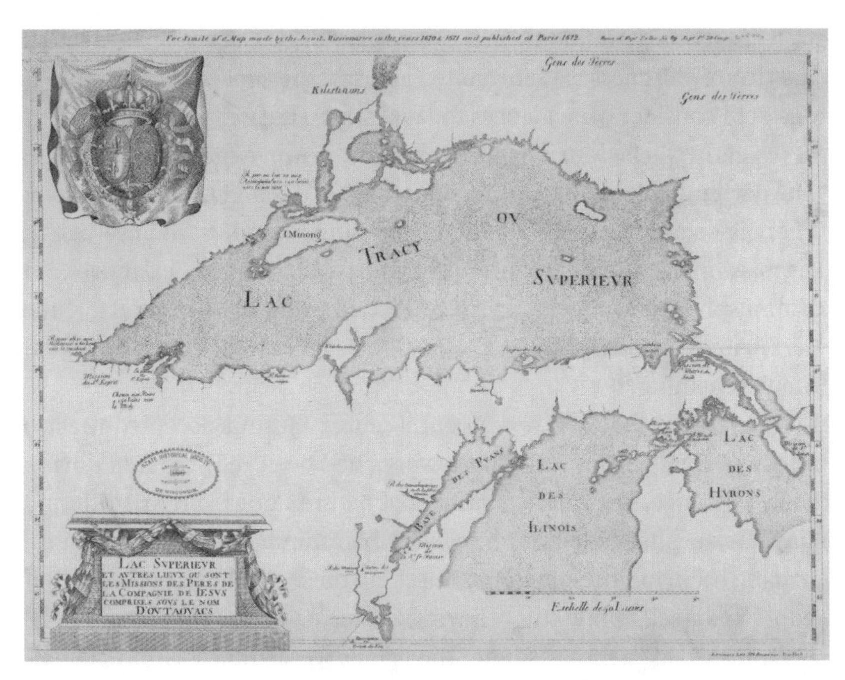

Father Claude Dablon, Superior-General of the Jesuit missions in New France, sketched this map after traveling the western Great Lakes in 1671. Father Gabriel Sagard served among the Huron in Thunder Bay, just beyond the right margin of the map, from 1623 to 1624. Louis Nicolas served at Mission St. Esprit at the extreme western end of the map, from 1667 to 1668. WHI IMAGE ID 39791

By 1640, they had ascended the St. Lawrence and fanned out through the eastern Great Lakes. During the 1650s, they entered the Iroquois country of western New York and Pennsylvania. In the 1660s, they reached Lake Superior and Green Bay and during the 1670s and 1680s paddled south through the Illinois country. In 1699, they reached the lower Mississippi Valley and a few years later established a church on the Gulf Coast at Biloxi, Mississippi. By the early eighteenth century, a chain of Jesuit missions stretched twenty-five hundred miles through the interior, from the Gulf of St. Lawrence in the northeast, westward through the Great Lakes, and south down the Mississippi to the Gulf of Mexico.[174] During those same decades, the methodology of modern science began to emerge. Sagard's naïve faith in Catholic dogma came under attack, and his successors found themselves caught between two opposing worldviews.

On the one hand, they felt certain that God and Satan were locked in an eternal combat unfolding in the physical world. They believed that the American wilderness was enchanted and saw the supernatural in things we would consider coincidences today. Father Paul Le Jeune, superior of the Jesuits in Quebec during the 1630s, told one American Indian shaman, "The devil meddles with your imaginations in the night." He also wrote: "There is no doubt that the Demons sometimes manifest themselves."[175] The missionaries lived, according to historian Dominque Deslandres, "in an atmosphere of wonders . . . in New France, every mark of good fortune was attributed to divine intervention," and every setback was credited to Satan's scheming.[176]

In scientific matters, seventeenth-century Jesuits accepted the authority of classical authors whose works had been rediscovered during the Renaissance. Their understanding of nature came from Aristotle and Pliny. These inherited ideas mingled in their minds with Christian faith, enabling them to simultaneously reject "all these silly ideas of these blind people [Indians]," while still believing in mermaids and unicorns.[177]

As absurd as that may sound, the Jesuits were known as "the schoolmasters of Europe," recognized for their wide knowledge and incisive minds. As the fledgling Scientific Revolution gained momentum during the middle of the seventeenth century, many Jesuits tried to embrace its principles—systematic observation, logical analysis, hypothesizing, testing, and concluding. Between 1620 and 1690, Jesuit missionaries produced more than eight hundred works on geography, zoology, botany, astronomy, anthropology, linguistics, and mathematics. Their center in Quebec even had a telescope, and the priests there astonished local tribes by predicting eclipses.

But as they spread out through the Americas, Asia, and Africa, the Jesuits saw birds that didn't fly, fish that soared out of the sea, hairless dogs that didn't bark, and other creatures that could not be explained by Aristotle or the Bible. This steady flow of new information was hard to reconcile with classical science or Catholic dogma. During the second half of the seventeenth century, Jesuit missionaries wrestled with these contradictions, as all Christians would two hundred years later after the revelations of Lyell and Darwin. One of the notable priests caught in this dilemma was Louis Nicolas (1634–ca. 1700), who wrote the first comprehensive work on North American birds.

Chequamegon Bay, where Jesuit missionaries Claude Allouez and Louis Nicolas served during the 1660s, from a nineteenth-century promoter's lithograph. WHI IMAGE ID 26099

Nicolas was born in 1634 in the south of France and began training as a Jesuit at age twenty. His teachers considered him "better suited for manual work and service than for intellectual activities," but after a decade of study and repeated requests, they allowed him to go to New France as a missionary in May 1664. He spent two years learning American Indian languages in Quebec (where his teachers found him "proficient at letters, weak in theology") before setting off, with Father Claude Allouez in August 1667, for Mission St. Esprit, on Lake Superior's Chequamegon Bay.[178]

Allouez had established this mission near modern Ashland, Wisconsin, in 1665 because hundreds of families from the Ottawa, Huron, Illinois, Ojibwe, and other nations had taken refuge there from Iroquois attacks farther east. Allouez wrote that they "cultivate fields of Indian corn and lead a settled life. They number eight hundred men bearing arms, but are gathered together from seven different nations, living in peace, mingled one with another."[179]

After settling Nicolas on the shore of Lake Superior, Allouez immediately departed for Green Bay, leaving his assistant to make converts all by himself. But Nicolas was more interested in worldly matters than spiritual ones. He spent much of his time traveling around the western Great Lakes studying wildlife and recruiting Indian hunters for French traders. Nicolas reportedly had "rough manners and behaviors" and probably fit right in with the uneducated voyageurs and fur traders.[180]

Nicolas was also pompous and arrogant. Allouez complained about "his lack of foresight in business, and his frequent and sudden movements of wrath." He behaved like a petty tyrant at Chequamegon Bay, trying to rule over local tribes rather than serve them. In 1669, an outraged Ottawa chief protested to the French governor that Nicolas had "carried these excesses so far as to beat himself, a chief of the nation, with a stick."[181]

In June 1668, Jesuit authorities called him back to Quebec, reprimanded him, and kept him under a close eye for two years before sending him to preach to the Mohawk in upstate New York. Again he spent much of his time wandering through the woods, sketching, and studying nature instead of making converts. In the spring of 1673, church authorities gave him one last chance and placed him in a remote outpost at the mouth of the St. Lawrence, but Nicolas disappointed them again and in the summer of 1674 they finally shipped him home to France.[182]

Nicolas brought back notes for three books. While detained in Quebec from 1671 to 1673, he had begun compiling a dictionary of Algonquian languages. In its introduction, he mentioned that he intended to follow it with two other books: one about North America's plants and animals and one about the indigenous peoples with whom he'd lived. They were to total twenty-four sections in three volumes, including a large series of illustrations that depicted Native peoples, plants, animals, and birds.[183]

With this work, Nicolas became the first writer to attempt to describe all the wildlife of North America. The Spanish had done it for Central American flora and fauna, and English writers had catalogued the natural history of the Caribbean and of Massachusetts Bay. But no one before Nicolas had tried to thoroughly describe the animals and plants living all across the northern tier of the continent, from Lake Superior to the Atlantic.[184]

"My God, how I regret embarking on an enterprise as difficult as this one," he wrote in about 1684. "What likelihood is there that, even after twenty years of assiduous work and repeated great travels, I can say all that is necessary about so many fine curiosities of a foreign country, where everything is different from ours?" He had charged himself with "treating in general and in particular: Simples [medicinal plants], flowers, grains, herbs, fruits, bushes, trees, four-footed animals living on land and in water, birds that live on land and those that live above or in water; and finally fresh-water fish, and some saltwater ones; various insects, and several reptiles, with their figures." He spent the rest of his life writing, revising, and sketching, and finally died without finishing his project.[185]

Nicolas left behind two manuscripts that were only published three hundred years later. His *Histoire Naturelle des Indes Occidentales* consists of 196 closely written pages measuring about thirteen by eight inches. Nicolas folded large sheets of blank paper and sewed them into notebooks that he filled with handwritten text describing 335 different animals and plants. Because it bears the ownership mark of a collector who died in 1689, this manuscript presumably dates from the 1680s. Today it is in the Bibliotheque Nationale and a digital facsimile is available online.

Nicolas also left a second manuscript of seventy-nine pages filled with drawings that was later given the title *Codex Canadensis*. Because most pages depict several different plants or animals, the album contains a total of 180 images. Besides flora and fauna, it includes two large maps, portraits of American Indian warriors or elders, and a few other sketches unrelated to the *Histoire Naturelle*. Nicolas's captions show that he was working on it as late as the year 1700 and that it was probably intended as a companion to his *Histoire Naturelle*. The *Codex* vanished after his death, only to resurface in the twentieth century. Today it is in the Gilcrease Museum in Tulsa, Oklahoma, and a full-color digital facsimile is available online.

Nicolas wrote more than fourteen thousand words about sixty-three birds in the *Histoire Naturelle* and drew fifty-six pen-and-ink sketches of them in the *Codex Canadensis*. This was by far the most thorough account of North American birdlife until Mark Catesby's *Natural History of Carolina, Florida, and the Bahama Islands* appeared in London between 1729 and 1747. Nicolas's drawings were the first in a tradition of illustrated works on North American birds that culminated one hundred fifty years later in Audubon's famous elephant folios.[186]

Unlike Sagard's books, which convey naïve trust and charming faith,

Owl, heron, and crane from Louis Nicolas's *Codex Canadensis*, ca. 1690.[187] GILCREASE MUSEUM, TULSA, OKLAHOMA

Nicolas's writings express the tension between religion and science that was transforming European intellectual life during the late seventeenth century. He prided himself on his empirical method: "I have made a careful study," he wrote in his introduction, "and I took great pleasure in observing everything beautiful and curious that I was able to see on my travels."[188] According to Nicolas, investigating closely was the mark of a good naturalist: "We cannot know or even understand other than by looking at [objects in nature] and examining them closely in the course of several repeated journeys, where it's necessary to study oneself, and to enjoy, or at least have the curiosity to look attentively at what one finds exceptional."[189] After mentioning "more than thirty writers" who had published natural histories, he insisted that "in [their] work I found nothing of which I put forward." He said he consulted them only "to learn about what I had not seen or where I had not been" personally.[190]

Despite this belief in the empirical method, Nicolas followed earlier models when classifying the birds that he observed. His *Histoire Naturelle* divides them into two sections, land birds and waterbirds, as every naturalist since Aristotle had done. And within each section Nicolas arranged birds by size, like all the naturalists who came before him: "Let us begin with the smallest and end with the largest, according to our custom."[191]

The smallest, of course, was the ruby-throated hummingbird, "the rarest, most beautiful and most wonderful of all those seen in the new world." Unlike some of his contemporaries, Nicolas did not mistake it for an insect. He wrote: "It flutters around like a bee, making a constant little pleasant buzzing, which is a little surprising when the bird passes by someone like lightning. It is so intent on feeding on the nectar of flowers that it can sometimes be caught easily, although it is always in the air. The noise that it makes with its wings prevents it from hearing the sound of someone approaching." When Europeans first discovered hummingbirds, he noted, "These birds were precious. Ladies of the court wore them instead of earrings; but the fashion has passed, and collectors are content to have them in their studies."[192]

Most previous observers wrote only about conspicuous large species or game birds, but Nicolas discussed nearly thirty small songbirds, too, and often related local beliefs about them. For example, he recorded that American Indians feared the whip-poor-will,

> in whose body the soul of a man or a woman enters, and comes and sings at the places where they camp . . . Our hunters say that it would not dare appear during the day, for people would make war against it. When they hear it, they greet the soul of the dead person who is in it, and they believe that it is one of their relatives: 'Kitaramikourimin tchipai-zen. We greet you, O soul of the dead.' And if the mood strikes someone, he immediately prepares food to feed this soul.[193]

As a devout Catholic, Nicolas considered such beliefs as heresies to be denounced and was quick to dismiss "the foolish ideas of our [Native] Americans concerning their beliefs about the dead."[194]

But to modern scientific eyes, Jesuit assumptions appear equally strange. For example, in addition to believing in mermaids and unicorns, the missionaries had unshakable faith in divine intervention. A colleague of Nicolas explained how a Jesuit priest tried to keep blackbirds from raiding a farmer's field with chants and incense:

> They enter'd the Field and walk'd through the Wheat in Procession, a young Lad going before the Jesuit with a Bason of their Holy-Water; then the Jesuit, with his Brush, dipping it into the Bason and sprinkling the Field on each side of him; next him, a little Bell tingling, and about thirty Men, following in order, Singing, with the Jesuit, Ora pro Nobis; at the End of the Field they Wheel'd to the Left about, and return'd. Thus they went through the Field of Wheat, the Birds rising before them and lighting behind them.

An English Protestant watching the ceremony noted: "At their return, I said to a French lad, The Fryar hath done no Service—He had better take a Gun and shoot the Birds. The Lad left me awhile (I tho't, to ask the Jesuit what to say) and when he returned, he said, the Sins of the People were so great, that the Fryar could not prevail against those Creatures."[195]

The missionaries also believed that some birds had been provided by God for human consumption. For example, Nicolas was astonished at the number of passenger pigeons in America and the ease with which they could be killed. In spring and fall, "For fifteen or twenty days one sees flocks pass which cover the sky all this time over a space more than three or four leagues long." They migrated in numbers "so great that people kill as many as they like with clubs in the streets of Mont Royal [Montreal]; they even shoot them from windows, when flocks of them are passing . . . seven to eight hundred of them [can be] taken by two or three persons, who stretch out a little net across a valley." He also noted that entire Indian villages relocated at pigeon-roosting places where young birds could be easily knocked off the nests: "Seven or eight thousand people, men, women and children, take part in this hunt for a month. They return so fat, along with their dogs, that it is unbelievable."[196]

The abundance of North American waterfowl also amazed Nicolas. Snow geese blanketed certain islands, turning them entirely white, and "It was no small marvel to see such great flocks of them that they covered the sun." Canada geese were also abundant. Lakes were

> so covered with them, that whenever one advances in a canoe or a rowboat, one hears what sounds like some horrible crash of thunder, accompanied by such loud repeated cries that I do not think an army of twenty or thirty thousand men would make more noise . . . With every stroke of the paddle we made . . . there rose so many flocks of bustards that someone less accustomed to this noise would have easily believed that he was in the middle of a battle."[197]

But the waterfowl that most impressed Nicolas was the tundra swan. He wrote:

> This is without doubt the king of the water birds. Its trumpet-like call, its proud bearing and majestic air, its admirable whiteness, its noble pride on the waters, which it seems to disdain when swimming, fleeing the tumult of other birds; all these make it noticeable above all other birds on the waters . . . But what I consider especially fine, is a certain down under the mantle of the swan, which is so beautiful and white that no more curious plumage could be seen. There is no ermine or snow whiter than the skin of this swan when it has been skinned and plucked to the down. If several of these skins were joined together after being well prepared by an Indian woman according to her custom, one could, without fear of being rejected, present them

> to a king, who would appreciate them because they are so rare and so useful. There is nothing more suitable to warm the chest of someone who has a chest cold, and to prevent whooping cough. For this reason the skins are valued and costly in France, where merchants bring as many as they can.[198]

Unlike modern biologists, Nicolas wasn't interested in nature for its own sake, and unlike Sagard, he didn't see animals and plants primarily as examples of God's glory. He focused instead on how nature was valuable to humans and was always quick to show some utilitarian purpose. For example, in his article on bald eagles he noted, "Some French people make fine candlesticks of the leg, the foot and the talons. It is amazing how fine and useful these candlesticks are when they are elegantly decorated with silver or some other metal, on which one places little candles of white wax. Many people have them as a curiosity in their chapels or their oratories."[199]

Although Nicolas attempted to employ the new empirical methods popularized during the Scientific Revolution, he was by no means scientific, as we would use the word today. "I know a Canadian Frenchman named Joliet," he wrote in his article on the golden eagle, "who between the age of six and seven was carried more than a hundred feet by an eagle that would have devoured him if he had not been rescued. I heard this from his own mouth and from those who had seen this terrible incident."[200]

But even a large golden eagle couldn't lift more than fifteen pounds. The irreverent fur traders who told this tale were probably pulling the leg of the priest known for his pomposity and arrogance. That Nicolas believed them reveals the limits of his science. He appealed to empirical evidence ("I heard this from his own mouth" and from "those who had seen this terrible incident"), but he didn't analyze the claim logically or test it the way a modern scientist would. Although he insisted on confirming many facts with his own eyes, Nicolas was also willing to accept whatever he learned from his Jesuit teachers, Aristotle, or even mischievous voyageurs.

In at least one instance, Nicolas may have gone beyond accepting another author's claims and actually plagiarized his words. Nicolas's discussion of the pelican closely resembles a passage in the *Jesuit Relation* of 1670 written by his colleague Claude Allouez. Nicolas's *Histoire Naturelle* says,

"From a distance, one would take them for swans, because their mantle and pennae [wing feathers] are all white, their neck is long." Allouez wrote, "One would take them for Swans, from a distance, as they have the latter's white plumage and long necks." Several other statements about pelicans in Nicolas's text are also strikingly similar to those of Allouez. It's probably impossible to say whether Allouez copied Nicolas's field notes in 1670 or whether Nicolas later paraphrased the 1670 *Relation* published in Paris.[201]

The only other eyewitness accounts of American birds available to Nicolas were Sagard's chapter and the *Histoire Véritable et Naturelle . . . du pays de la Nouvelle France* (Paris, 1664) by Pierre Boucher, governor of Quebec, and there's no evidence that he relied significantly on either work. Boucher's chapter on birds is a fraction of the length of Nicolas's, confines itself mainly to birds of eastern Quebec, and contains very few observations made in the field. When the two authors' treatments of the same birds are compared side by side, there's no evidence that either one was indebted to the other.

When preparing his drawings, on the other hand, Nicolas often followed well-known models. For example, his image of the pelican on page 54 of the *Codex* shows the bird gobbling up a fish in exactly the same manner that a century-old drawing by Conrad Gesner did, and his figures of American Indians often mirrored those of his contemporary, Francois Du Creux.[202]

Despite their intellectual richness and obvious charm, Nicolas's writings and drawings were not published for more than three hundred years. Before a book could be printed in France in the seventeenth century, Catholic officials had to certify that it contained no heresies and government censors had to approve it. Nicolas's manuscript on plants and animals shows marks that may have been made by the censors, and in one place he says that he's revised his text to please certain readers.

However, church authorities were not pleased. They may have objected to his praise of Native peoples, who were nothing more than ignorant heathens in the eyes of the church, or they may have been prejudiced against Nicolas because of the troubles he had caused in America. Whatever the reason, Jesuit officials refused to approve printing his proposed works on North America. He protested their decision in a letter dated November 10, 1677, and a year later he left the Society of Jesus altogether. Whether

Nicolas resigned from the order or was expelled is unclear, but the separation was probably welcomed by both parties.[203]

After the Jesuits refused to allow his book to be published, Nicolas probably decided to create a single manuscript copy to be lavishly bound and presented to a patron. This was a common practice at the time, the uniqueness of the volume adding value to the gift. Like other well-traveled Jesuits, Nicolas had access to the French court and may even have intended to present a final, perfected copy to a member of the royal family.[204]

He may have given *Histoire Naturelle* to Pierre de Maridat, a counselor to King Louis XIV whose ownership marks appear on it. Maridat died in 1689 and the manuscript resurfaced again, in the Bibliotheque Nationale, after the Revolution of 1789. The *Codex Canadensis* disappeared for 250 years until it suddenly showed up in 1930 in the hands of a Paris book dealer. After passing briefly through several private owners, it was bought by American collector Thomas Gilcrease in 1949.[205]

Nicolas worked on the *Histoire Naturelle* and *Codex Canadensis* for most of his adult life. From 1664 to 1675, he explored the forests and streams of North America jotting notes, making sketches, and assembling drafts. From 1676 into the 1680s, he "corrected all [his] papers many times word by word."[206] For another decade he perfected and captioned his drawings. Then, instead of reaching a wide popular audience like other contemporary books written by visitors to America, his lifework vanished. Internal evidence proves that he was revising the *Codex* as late as the year 1700, after which literally nothing is known about him. Even the date of his death and his final resting place are unknown.

Many other missionaries recorded impressions of American wildlife at the end of the seventeenth century that did not vanish, but were widely read. The Jesuit order required them to contribute letters and reports to its international scholarly network. "When one leaves France for distant countries," one Illinois missionary wrote home, "it is not difficult to make promises to one's friends; but, when the time comes, it is no slight task to keep them, especially during the first years. We have here but a single opportunity, once a year, for sending our letters to France. It is therefore necessary to devote an entire week to writing, without interruption, if one wishes to fulfill all one's promises."[207]

In addition to personal letters, missionaries throughout the Great

Lakes and Mississippi Valley sent annual reports to their superior in Quebec, who edited them each year into a single volume. These were published annually in Paris from 1632 to 1673 under the title *Relation de ce qui s'est passé de plus remarquable aux missions des peres de la Compagnie de Jesus en la Nouvelle France les années. . .* (*Account of the most remarkable things that happened in the Jesuit mission in New France during the years. . .*). The modern bilingual edition of the *Jesuit Relations*, as they are usually called, fills seventy-three volumes. They contain a wealth of unparalleled eyewitness accounts of life in the Midwest just after the arrival of Europeans and are dotted with observations of American birds.[208]

Apart from Sagard and Nicolas, most missionaries made only passing references to birdlife. In a typical comment, Claude Allouez wrote from Green Bay in April 1670: "We slept at the head of the bay, at the mouth of the River des Puans . . . On our way we saw clouds of Swans, Bustards, and Ducks."[209] Jacques Marquette, whom Allouez brought out to Lake Superior to replace Nicolas, was equally terse: "There is fine hunting there [in Illinois] of Wild Cattle, Bears, Stags, Turkeys, Ducks, Bustards, Pigeons, and Cranes." Having cited birds' economic value as food, most missionaries had said as much as they cared to about Midwestern avian life.[210]

The Jesuits frequently speculated on whether American birds were the same as those in France. Sebastien Rasles compared French and Illinois turkeys in a letter to a friend back home: "We can hardly travel a league without meeting a prodigious multitude of Turkeys, which go in troops, sometimes to the number of 200. They are larger than those that are seen in France [which are, in fact, an entirely different species]. I had the curiosity to weigh one of them, and it weighed thirty-six livres [about forty pounds]."[211]

Louis Vivier compared the taste of an unidentified American game bird to those of France: "There are also certain birds as large as hens, which are called pheasants in this country, but which I would rather name 'grouse;' they are not, however, equal in my opinion to the European grouse. I speak not of partridges or of hares, because no one condescends to shoot at them."[212]

In 1670, Marquette learned from Illinois Indians visiting Chequamegon Bay that a vast populated country lay south of Lake Superior, and he asked to preach to the tribes there. About the same time, French author-

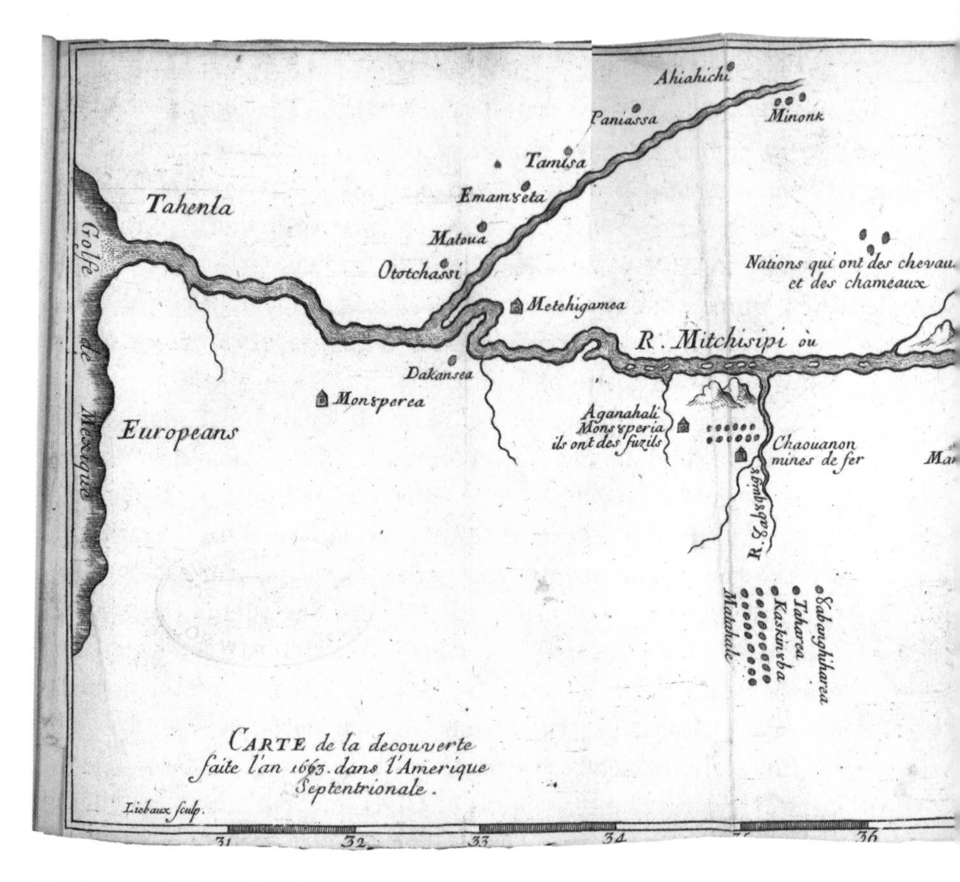

The first map of the entire Mississippi River, made from Marquette and Joliet's notes and published in 1681. Green Bay is at the bottom right and New Orleans at the upper left.
WISCONSIN HISTORICAL SOCIETY LIBRARY, DIGITAL IDENTIFIER TP103001

ities heard reports from traders of an immense river in that direction that might provide a water route through the entire continent. Early in 1673, the governor of New France ordered twenty-seven-year-old Louis Joliet, a former philosophy student who had taken up fur trading, to find out whether the river went to the Pacific Ocean or perhaps to the hypothetical gold mines of Quivira, for which the Spanish had crossed the Great Plains more than a century earlier. They allowed Marquette to go with Joliet and preach to tribes along the way.[213]

Joliet was an experienced backwoods traveler who knew native languages and possessed "the Courage to dread nothing where everything

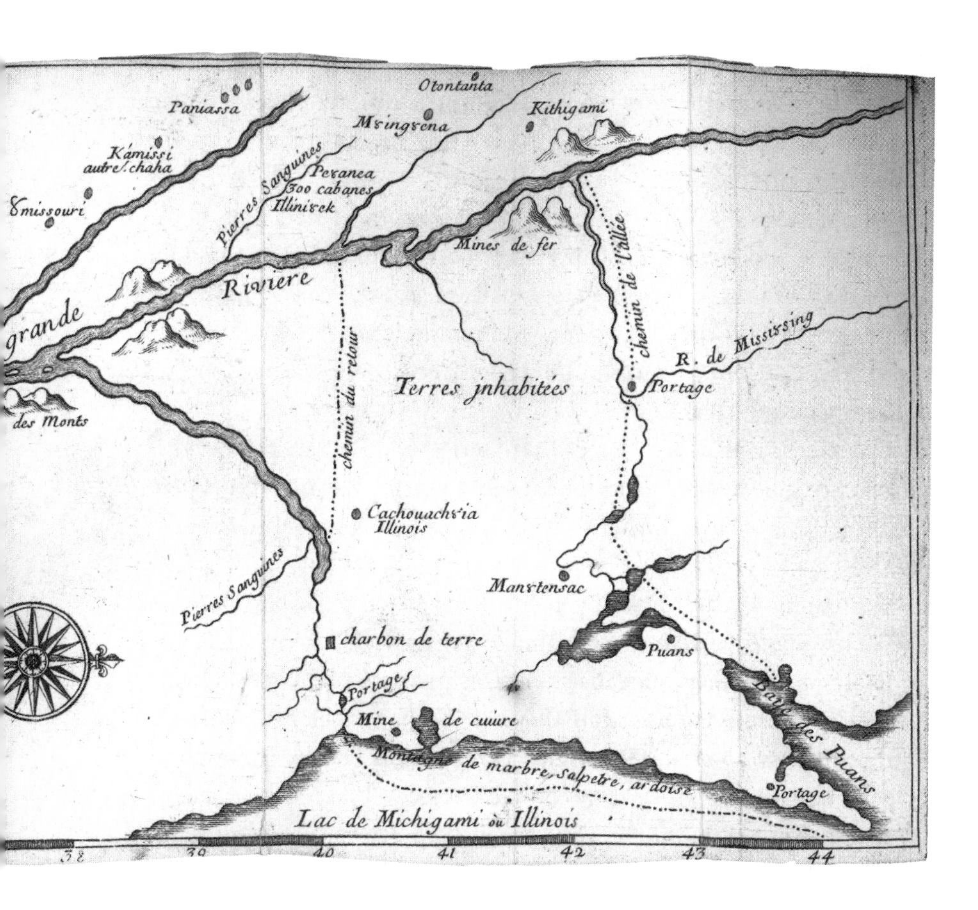

is to be Feared." Marquette had tact, patience, humility, and intellectual curiosity. Their twenty-five-hundred-mile canoe trip in 1673 planted the seeds that later sprouted the first white settlements in the Mississippi Valley, introduced Christianity into six hundred thousand square miles of wilderness, and ultimately gave French names to cities from Eau Claire, Wisconsin, to New Orleans, Louisiana. They also recorded the first information about birds in the Mississippi Valley.[214]

Joliet and Marquette left Mission de St. Ignace, where Lake Huron meets Lake Michigan, on May 17, 1673. Heading southwest through Green Bay, Marquette in good Jesuit fashion investigated whether Lake Michigan was subject to tides (he concluded it was). They headed up the Fox River, which was "full of bustards, Ducks, Teal, and other birds, attracted thither by the wild oats, of which they are very fond," and then down the Wisconsin River to the Mississippi. Paddling south, they "saw only deer and cattle, bustards, and Swans without wings, because they drop Their plumage in This country."[215]

Marquette and Joliet made notes about everything new that they encountered. Louis Nicolas probably knew both men, and they could have inspired a fabulous monster he drew in the *Codex*. "We saw on the water a monster with the head of a tiger," Marquette wrote, "a sharp nose like that of a wildcat, with whiskers and straight, erect ears; the head was gray and the neck quite black."[216]

One pictograph of a monster Marquette and Joliet encountered was mistakenly called a bird for three hundred years. They saw it just before reaching the mouth of the Missouri River near modern Alton, Illinois:

> Skirting some rocks, which by their height and length inspired awe, We saw upon one of them two painted monsters which at first made us afraid, and upon which the boldest savages dare not long rest their eyes. They are as large as a calf; they have horns on their heads like those of deer, a horrible look, red eyes, a beard like a tiger's, a face somewhat like a man's, a body covered with scales, and so long a tail that it winds all around the body, passing above the head and going back between the legs, ending in a fish's tail. Green, red, and black are the three colors composing the picture. Moreover, these 2 monsters are so well painted that we cannot believe that any savage is their author; for good painters in France would find it difficult to paint so well,—and, besides, they are so high up on the rock that it is difficult to reach that place conveniently to paint them.

A sketch that Joliet made apparently reached French cartographer Jean Baptiste Louis Franquelin, who included it as a decoration on his 1678 manuscript map, "Carte de la France septentrionalle: contenant la decouverte du pays des Illinois." The monstrous pictographs were noted by other early explorers until 1699, when a passing missionary found them "now almost effaced."[217]

Franquelin's image from Joliet's sketch shows a creature without wings. Historian Natalia Belting pointed out that it appears to be not a bird but rather a water panther from the Lower World, "one of the most common symbols of evil in the cosmogony of the Algonquian and Siouan Indians of the Mississippi Valley and the Great Plains." In the early nineteenth century, white residents painted a new image on nearby rocks that loosely resembled a European dragon. They gave it wings and called it the Piasa Bird. Later settlers invented imaginary American Indian stories about it and called it a Thunderbird. Marquette and Joliet were then said, incorrectly, to be the first white men to see an image of a Thunderbird, when in fact they saw its cosmological opposite.[218]

When Marquette and Joliet reached the southern Mississippi Valley, they "saw Quail on the water's edge. We killed a little parroquet, one half of whose head was red, the other half and the neck yellow, and the whole body green." Unlike the water tiger, this was not a figment of Marquette's imagination but North America's only native parrot. The Carolina Parakeet was about the size of a grackle, and its gaudy colors, screeching call, and gregarious habits caught the eye of European observers for nearly three hundred years, until the last one died in captivity in 1918.[219]

Joliet and Marquette turned back near modern Helena, Arkansas, after learning that they were just a few days from the sea and approaching Spanish-controlled territory. They had proved that the Mississippi did not flow into the Pacific, but it could carry trade from the Great Lakes to the Gulf of Mexico.

When they got back to Lake Michigan, Joliet headed east to deliver their journals and specimens to officials in Quebec while Marquette went down to Illinois to preach to local tribes. Joliet nearly perished when his canoe capsized in the Lachine Rapids outside Montreal. His diary, his specimens, and his crew were all lost, but he said: "I was saved after having spent four hours in the water, having lost consciousness, by some fisherman, who never go to this strait and who would not have been there if the Blessed Virgin had not obtained for me this grace from God, who stayed the course of nature in order to save me from death." Marquette, his health weakened by exposure, malnutrition, and disease from the trip with Joliet, died the next year in the wilderness.[220]

Just as Joliet credited his rescue to divine intervention, missionaries in the Great Lakes and Mississippi Valley always saw the hand of God at work when they described American birds. For example, Louis Hennepin, a priest of the Recollect order who traveled the Upper Mississippi Valley with the explorer La Salle in about 1680, recalled how God intervened to save him:

> 'Twas not every day we met with any game, nor when we did, were we sure to kill it. The eagles, which are to be seen in abundance in these vast countries, will sometimes drop a breme, a large carp, or some other fish, as they are carrying them to their nests in their talons, to their young. During this scarcity of food He who cares for the smallest birds caused us to see crows and eagles gathered on the lakeshore. Paddling with redoubled effort toward these carnivorous birds, we found half of a very fat deer which wolves had strangled and partly consumed. We renewed our strength with meat from this animal, blessing Providence which had sent such timely aid.[221]

Although modern science had been born, it was still in its infancy, and the first white people to write about Midwestern birds had not mastered the scientific method. Instead, they lived in a universe populated by supernatural forces. It would be another century before science took firm root in the minds of observers writing about American birds.

6

Soldiers, Statesmen, and Science

Today, common sense tells me that the world is filled with discrete objects, that they're composed of molecules and atoms, that events in the world happen randomly (apart from the laws of physics), and that an object's value depends mainly on the price for which it can be sold.

But 350 years ago, common sense told intellectuals like Marquette, Nicolas, and Sagard that the world was enchanted, that objects were infused with spiritual significance, that events happened according to God's will, and that an object's value depended on its place in a Great Chain of Being with God at the top and angels, souls, humans, animals, plants, and minerals ranked below Him.[222]

The shift from their worldview to ours took place very gradually as new discoveries and educational reforms replaced old assumptions with novel ones. In 1600, astronomer Giordano Bruno was burned at the stake in Rome for insisting that the sun was just one of many stars and the earth just one of many planets. But in 1700, after another century of observations with the telescope, that idea was widely accepted by educated people. During the course of the eighteenth century, experiment and observation slowly replaced inherited dogma as the standard of intellectual authority. Reason and science superseded faith and magic in the minds of people who wrote about birds.

The records of Marquette and Joliet's expedition, like the notebooks of Louis Nicolas, vanished for nearly two hundred years. In 1672, the Pope halted publication of the *Jesuit Relations*, and the volume that would have included their voyage never appeared. But the superior of missions in

New France interviewed Joliet, prepared a summary of the voyage from Marquette's journal, and sent that document home to France for private circulation. Officials in Montreal and Paris immediately saw that they could travel safely by water from the Atlantic to the Gulf of Mexico. Possibilities for exploiting the riches of the interior—mostly furs—seemed limitless. These officials wanted to keep this knowledge from rival British and Spanish explorers, but it was leaked to the world eight years later in a collection of travels printed in Paris.[223]

Among those intrigued by the discoveries in America's heartland was thirty-year-old Rene-Robert Cavelier de La Salle (1643–1687). He had been educated by the Jesuits and taught school in France before joining an uncle in Montreal in 1667. Realizing that Marquette and Joliet had opened the door to a huge territory full of untapped resources, he secured permission from French authorities in 1678 to explore and colonize the Mississippi Valley. He was allowed to finance his explorations through a monopoly on the fur trade between Lake Erie and the Gulf of Mexico. Between 1679 and 1684, La Salle led six expeditions of soldiers, traders, and laborers thousands of miles through the interior to erect forts, negotiate treaties with American Indian nations, and open trading posts.[224]

La Salle's own writings and the amazing memoirs of his subordinates contain some of the finest early descriptions of birds in the Mississippi Valley. Their first concern was not birds, of course, but how to stay alive and establish base camps. They had no time, either, for scientific observations or religious musings. Birds were important to them mainly as a food supply and as a propaganda device to persuade supporters back in France that their work was successful and important.

One of La Salle's soldiers, Captain Pierre-Charles de Liette, who lived with the Peoria and Kaskaskia tribes on the Illinois River for fifteen years (1687–1702), commented often on birds. On the topic of waterfowl migrations, de Liette wrote:

> I am now going to relate something which will perhaps not be believed, though I am not the only one who has witnessed it. The waters are sometimes low in autumn so that all the kinds of birds I have just mentioned leave the marshes which are dry, and there is such a vast number of them in the river, and especially in the lake (at the end

> of which the Illinois are settled on the north shore), on account of the abundance of roots in it, that if they remained on the water, one could not get through in a canoe without pushing them aside with the paddle, and yet the lake is seven leagues long and more than a quarter of a league wide in the broadest part [about twenty-four miles long and a mile wide].[225]

In a letter dated October 8, 1701, soldier and trader Antoine de Lamothe Cadillac praised Detroit's natural riches in similarly generous terms:

> It is along these vast trails that we see congregated by hundreds the timid deer and shrinking doe, with the buck bounding eagerly to gather the apples and plums with which the ground is paved; it's there that the watchful turkey calls together her numerous brood to harvest grapes; it's there that their males go to satisfy themselves. The golden pheasants, the quail and partridge, the abundant turtle dove swarm in the woods and over the fields broken by clusters of tall forest trees, which afford a charming prospect such as alone can assuage the loneliness of solitude . . . The swans are so great in number, that one might mistake them for water lilies among which they are entangled.[226]

Hyperbole like this was useful for building political and financial support back home, but it was deliberately exaggerated. La Salle and his lieutenants rarely mentioned that the astounding abundance of waterfowl was seasonal and that at other times of year they nearly starved to death for lack of game.

In one instance, a turkey inadvertently aided La Salle in gaining crucial strategic information. Trekking alone across Illinois in the winter of 1680,

> he was some two leagues from the fort, carrying four turkeys which he had killed [when] he met a young Illinois warrior—one of a party returning with the prisoners from the south—who had hurried on in advance of the rest in order to announce their return at the village. Being very tired and hungry he begged food of M. de la Salle, who gave him one of his turkeys. Lighting a fire, the Savage put the turkey

> in the kettle he was carrying. While it was cooking, M. de La Salle questioned him about his journey, asking news from the lower river country, with which he pretended to have considerable acquaintance. With a bit of charcoal the young man drew upon a sheet of bark a very accurate map, stating that he had been everywhere in his pirogue, that one could reach the sea without encountering either falls or rapids, but that, where the river became very wide, there were occasional sand bars and mud banks, which would choke some part of the channel. He also gave the names of the tribes living along its banks, and the names of its tributaries.[227]

Equipped with this information, La Salle continued south in pursuit of his vision of a connected series of trading posts arcing across the entire heart of the continent.

In 1684, he sailed with three hundred colonists from France to the Gulf of Mexico, intending to populate the Illinois country by coming up the Mississippi River. Unfortunately, they overshot the river's mouth and were shipwrecked in Texas, where most of the expedition slowly died of starvation and exposure. After three years of fruitless wandering in Texas and Louisiana, a mutinous soldier finally assassinated La Salle. A handful of surviving colonists then set out overland and eventually reached Michilimackinac, at the head of Lake Michigan. There they found Baron Louis-Armand de Lahontan, a twenty-two-year-old lieutenant seeking his fortune in the New World, who left the next account of Midwestern birds.[228]

Lahontan (1666–ca. 1716) came to Canada as a teenager in 1684 in pursuit of a military career. By age twenty-two, he had fought the Iroquois in western New York, commanded a French fort at modern Port Huron, Michigan, and accompanied Indian hunters on expeditions throughout the Great Lakes. Four months after greeting La Salle's survivors in May 1688, he crossed Lake Michigan, traversed the Fox–Wisconsin waterway, paddled north up the Mississippi, and then (so he claimed) traveled west for hundreds of miles up a "Long River" toward the Pacific. It's impossible to say precisely where Lahontan went or what he saw that winter, or if he traveled west at all; his account of a western trip may have been simply a fiction inserted into his book for political reasons. He returned to the Mis-

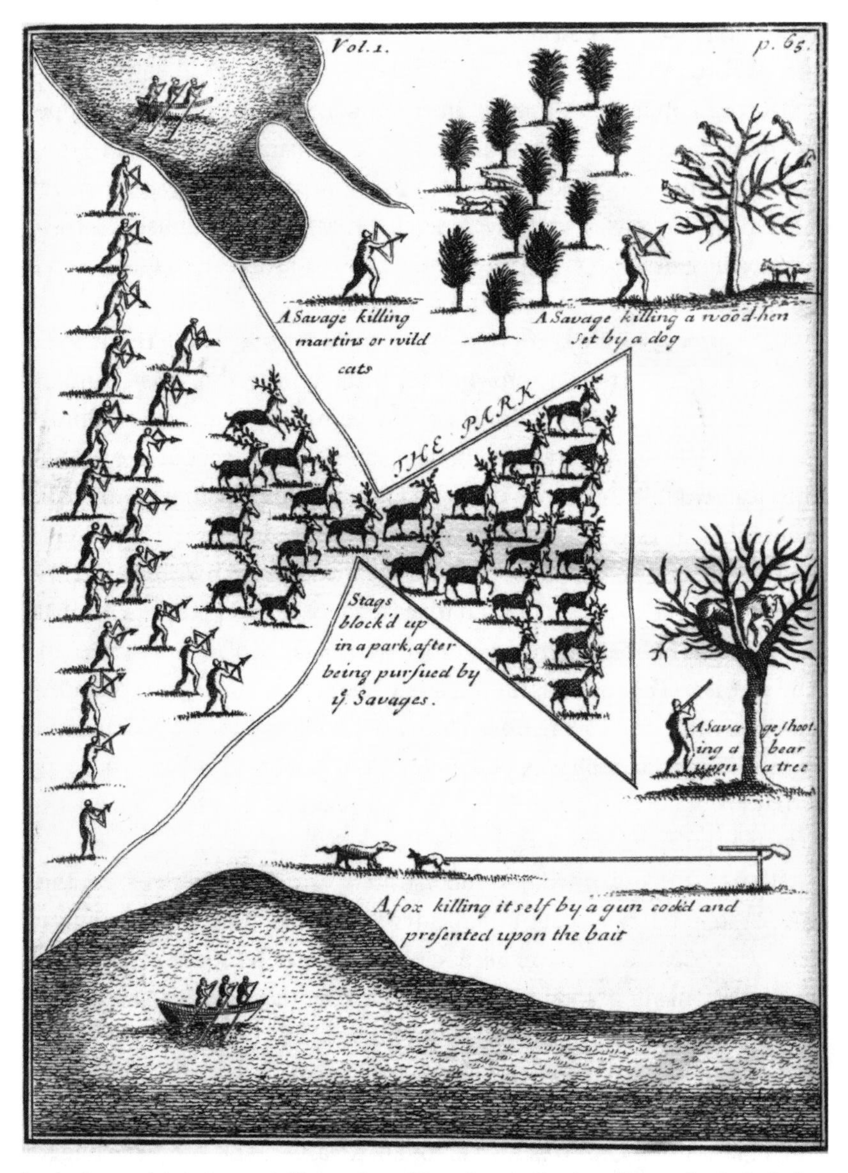

Louis-Armand de Lahontan's illustration of American Indian hunting methods, including bow and arrow, musket, trap, and enclosure, published in London in 1703. WISCONSIN HISTORICAL SOCIETY LIBRARY, DIGITAL IDENTIFIER AJ-145-0015

sissippi in March of 1689 and was back in Montreal the next year fighting English invaders.

During Lahontan's absence, his political enemies at home stripped him of his family fortune and undermined his backing at court. In 1693, he deserted his command and returned to Europe, where he spent the rest of his life trying unsuccessfully to regain his estates. He published a travel book, *Nouveaux Voyages . . . dans l'Amerique Septentrionale,* in 1703, quickly followed by his *Memoires* later the same year.[229]

Lahontan's flamboyant prose and colorful stories made these books best sellers, and translations and reprints appeared almost annually throughout the opening decades of the eighteenth century. Although he was not a naturalist and did not, like Louis Nicolas, set out to catalogue American wildlife, his books contained the most thorough eyewitness accounts of Midwestern birds for two generations.

The Age of Reason had begun and skepticism was in fashion in Paris, so Lahontan had to justify himself to his readers in strong terms. "All that I writ in the foregoing Letters, and the whole substance of the Memoirs I now send you, is truth as plain as the Sunshine," he claimed in his 1703 preface. "I flatter no Man, and I spare no body. I sacrifice all to the love of Truth, and write with no other view, than to give you a just Representation of things as they are . . . In the course of my Voyages and Travels, I took care to keep particular Journals of every thing."[230]

He poked holes in medieval superstitions, criticized the clergy for their naïve faith in religious dogma, and advocated clear thinking and common sense. He observed American birds closely, even sometimes providing precise measurements of specimens, and devoted one section of his *New Voyages* entirely to birdlife. In it he listed more than fifty species of Midwestern birds, about a dozen of which he described in substantial detail.

Lahontan's passage about unexpected winter invasions of the willow ptarmigan nicely captures his personality, values, and literary style (or perhaps that of his eighteenth-century London translator):

> The White Partridges are as big as our red Partridges, Their Feet are covered with such a thick down, that they resemble those of a young Rabbet. They are only seen in the Winter time, and some years they are scarce seen at all, though on the other hand, in other years they are

> so plentiful, that you may buy a dozen for Nine pence. This is the most stupid Animal in the World; it sits upon the Snow, and suffers it self to be knock'd on the head with a pole without offering to stir. I am of the opinion, that this unaccountable numbness is occasion'd by its long flight from Greenland to Canada. This conjecture is not altogether groundless, for 'tis observ'd, that they never come in flocks to Canada but after the long continuance of a North or a North-East Wind.[231]

His description of the ruby-throated hummingbird is one of the more detailed and accurate early accounts:

> The Flylike Bird is no bigger than one's Thumb, and the colour of its Feathers is so changeable, that 'tis hard to fasten any one colour upon it. They appear sometimes red, sometimes of a Gold colour, at other times they are blew and red; and properly speaking, 'tis only the brightness of the Sun that makes us unsensible of the change of its gold and red colours. Its beak is as sharp as a Needle. It flies from Flower to Flower, like a Bee, and by its fluttering sucks the flowery Sap. Sometimes about Noon it pearches upon the little branches of Plum-trees or Cherry-trees. I have sent some of 'em dead to France, it being impossible to keep 'em alive, and they were look'd upon as a great Curiosity.[232]

The number of migrating passenger pigeons astounded the young nobleman. While with an Indian hunting party in upstate New York, Lahontan and his guides

> resolv'd to declare War against the Turtle-Doves, which are so numerous in Canada that the Bishop has been forc'd to excommunicate 'em oftner than once, upon the account of the Damage they do to the Product of the Earth. With that view, we imbarqued and made towards a Meadow, in the Neighbourhood of which, the Trees were cover'd with that sort of Fowl, more than with Leaves: For just then 'twas the season in which they retire from the North Countries, and repair to the Southern Climates; and one would have thought, that all the Turtle-Doves upon Earth had chose to pass thro' this place. For the eighteen or twenty days that we stay'd there, I firmly believe that

> a thousand Men might have fed upon 'em heartily, without putting themselves to any trouble.[233]

In similarly colorful language, Lahontan describes American Indians hunting loons by canoe for sport, eagles and osprey contesting in the air for fish, and wood ducks whose "Feathers upon their Neck looks so bright, by vertue of the variety and liveliness of their colours, that a Fur of that nature would be invaluable in Muscovy or Turky." He distinguished the American bluebird from the European nightingale in being "of a lesser size than the European and of a blewish colour, and its notes are more diversified; besides that, it lodges in the holes of Trees, and four or five of 'em do commonly keep together upon the thickest Trees, and with joynt Notes Warble o'er their Songs."[234]

During the decades that followed Lahontan's brief stay in North America, French settlers gradually transformed La Salle's seasonal trading posts into permanent towns. The cities of Detroit, Vincennes, Green Bay, Prairie du Chien, Kaskaskia, and St. Louis were all founded by French settlers by 1765. They were often populated by aging fur traders and voyageurs who could no longer endure the hardships of the wilderness, but preferred an unfettered life on the frontier to the social strictures of civilization in Montreal or Quebec. Many married Indian women, and interracial families were ubiquitous throughout the region.

Few of these eighteenth-century residents of the interior were sufficiently literate to leave any written record of the natural world. But between 1755 and 1763, when French and British soldiers fought for control of North America, visiting military officers wrote accounts of the interior that included descriptions of birds. By then, magic and mystery had largely given way to reason and science in the minds of educated Europeans. Scientific expeditions such as those of Captain James Cook and the Comte de La Pérouse would soon be sent into remote regions at immense expense, merely to gather facts and expand knowledge.

The flow of new biological data coming into Europe from explorations in Africa, Asia, and the Americas was systematized by the Swedish naturalist Linnaeus (1707–1778), who invented the first comprehensive classification for all living organisms and fixed their names unambiguously. His first attempt to organize the wave of new data appeared in

1735, and for two decades, as colleagues, correspondents, and students sent him specimens from around the globe, Linnaeus issued updated editions of his classification scheme, *Systema Naturae*. He advocated that living things should be designated by two sequential Latin terms, one placing it in a genus and the other describing its species, and he provided rules for proper classification and naming. These were first applied to birds in the tenth edition of *Systema Naturae*, which appeared in 1758.

Linnaeus's ideas reached America quickly. Cadwallader Colden (1688–1776) could buy a copy of a Linnaean botanical work in New York in 1739, and in 1748 Linnaeus's student Pehr Kalm was greeted on his arrival by Benjamin Franklin, who was well informed about his mentor's work. Franklin put Kalm in touch with virtually every important naturalist in the United States, and Kalm urged them all to adopt Linnaean naming conventions. The Linnaean bird names also seeped into popular discourse through English travel books, gardening manuals, periodicals, and encyclopedias that were exported to the colonies.

By the time Kalm returned to Europe in 1751, France and Britain were on the brink of a war for the eastern half of North America. This seven-year conflict brought thousands of British soldiers into the Midwest for the first time. They recorded the first words in English about the heart of the continent, including the region's birds. Unlike the Jesuit missionaries discussed in the previous chapter, they generally saw America through the lens of science.

For example, James Smith, a British soldier taken prisoner by a band of Ottawa Indians near Lake Erie in November 1756, described one of his captors' beliefs about natural history:

> It is a received opinion among the Indians, that the geese turn to beavers and the snakes to raccoons; and though Tecaughretanego, who was a wise man, was not fully persuaded that this was true; yet he seemed in some measure to be carried away with this whimsical notion. He said that this pond had been always a plentiful place of beaver. Though he said he knew them to be frequently all killed, (as he thought;) yet the next winter they would be as plenty as ever. And as the beaver was an animal that did not travel by land, and there being no water communication, to, or from this pond—how could such a

> number of beavers get there year after year? But as this pond was also a considerable place for geese, when they came in the fall from the north, and alighted in this pond, they turned beavers, all but the feet, which remained nearly the same.

Smith thought this was nonsense, and when he offered an alternative explanation for how the beaver might survive the winter, he was pleased that his host "granted that it might be so."[235]

Smith also related how the Ottawa explained noisy Canada geese flocking in nearby marshes as winter approached:

> The geese at this time appeared to be preparing to move southward—It might be asked what is meant by the geese preparing to move? The Indians represent them as holding a great council at this time concerning the weather in order to conclude upon a day, that they may all at or near one time leave the Northern Lakes, and wing their way to the southern bays. When matters are brought to a conclusion and the time appointed that they are to take wing, then they say, a great number of expresses are sent off, in order to let the different tribes know the result of this council, that they may be all in readiness to move at the time appointed. As there is a great commotion among the geese at this time, it would appear by their actions, that such a council had been held.

This seemed more plausible to Smith than geese turning into beavers. "Certain it is," he said, "that they are led by instinct to act in concert and to move off regularly after their leaders."[236]

Jonathan Carver of Massachusetts, who, like Smith, was a soldier in the French and Indian War, made similar appeals to reason after traveling in the Upper Mississippi Valley on an unsuccessful quest for a northwest passage. From August 1766 to the spring of 1767, he journeyed from Michilimackinac through Green Bay and across Wisconsin, then north up the Mississippi and onto the Minnesota prairies, where he met with Sioux leaders. His memoir of the trip, *Travels through the Interior Parts of North America in the Years 1766, 1767, and 1768*, provided the most thorough report on Midwestern birds since Lahontan in 1703.[237]

Carver listed forty-one species in his opening paragraph and referred to another "upwards of twenty species" of wild ducks. His purpose was "in the first place, [to] give a catalogue, and afterwards a description of such only as are either peculiar to this country, or which differ in some material point from those that are to be met with in other realms." These numbered some eighteen birds, to which he devoted ten pages.[238]

Carver felt compelled to apologize for not always meeting the standards of science. "I am sensible," he wrote, "that I have not treated the foregoing Account of the natural productions of the interior parts of North America with the precision of a naturalist. I have neither enumerated the whole of the trees, shrubs, plants, herbs, &c. that it produces, nor have I divided them into classes according to their different genera after the Linnæan method." He blamed this in part on "the limits of my Work, [which] in its present state, would not permit me to pursue the subject more copiously" and pledged that if enough subscribers for the book came forward, he would issue an enlarged edition with "many interesting particulars and descriptions, which the size of the present Edition obliges me to curtail or entirely to omit."[239]

But Carver's education before taking up the pen had been in a small-town school, and his only known professions were shoemaking and soldiering, so it's unlikely that he could ever have written like a university-educated naturalist. He understood, however, that he had to fulfill readers' scientific expectations and took care to defend his methods with phrases such as, "This conclusion is the most rational I am able to draw . . . ," "To me it appears highly improbable that . . . ," and "the limits of my present undertaking will not permit me . . . to enumerate any other proofs in favour of my hypothesis."[240] Although he lacked their training, Carver employed the precise observational methods and skepticism of contemporary scientists. He meticulously described how eagles scavenged below the Falls of St. Anthony in modern St. Paul, for example; how "exceedingly nimble" loons evaded Indian hunters; and the intricate shades of plumage on the blue jay.

Carver also tried to employ scientific skepticism, though he was not always successful. The osprey, he wrote, "sometimes seems to lie expanded on the water, as he hovers so close to it, and having by some attractive power drawn the fish within its reach, darts suddenly upon them." He then reported the popular explanation for this behavior: "The charm it makes use of is supposed to be an oil contained in a small bag in the body, and which nature has by some means or other supplied him with the power of using for this purpose." Though dubious about this explanation, Carver nevertheless passed along the misconception that "any bait touched with a drop of the oil collected from this bird is an irresistible lure for all sorts of fish, and insures the angler great success." He was enough of a scientist to question, but not to test or reject, the popular fallacy.[241]

Carver reported other beliefs of local residents, even when he doubted them. He wrote of the whip-poor-will, "As soon as night comes on, these birds will place themselves on the fences, stumps, or stones that lie near some house, and repeat their melancholy notes without any variation till midnight. The Indians, and some of the inhabitants of the back settlements, think if this bird perches on any house, that it betokens some mishap to the inhabitants of it."[242]

By the time Carver's book appeared in 1778, the American Revolution—which he called "the unhappy divisions that at present subsist between Great Britain and America"—was well under way.[243] Fighting against England in an American regiment was twenty-four-year-old Gilbert Imlay (1754–1828), another soldier who wrote at length about Midwestern birds.

Imlay was born in New Jersey, was commissioned a lieutenant during the Revolution, and went west to seek his fortune after the war. A few years later, he fled to England to escape unpaid debts and prosecution for unscrupulous business practices that included cheating Daniel Boone in a real estate deal. In England, he became the lover of feminist pioneer Mary Wollstonecraft, moved in a circle of revolutionary thinkers and romantic writers, and in 1792 published *A Topographical Description of the Western Territory of North America*, which includes many references to the birds of the Ohio Valley.[244]

Imlay's book purports to be a series of letters from Kentucky intended to draw attention to the untapped potential of the Midwest, so he uses birdlife as a symbol of abundance:

> The rapidity of the settlement has driven the wild turkey quite out of the middle countries [mid-Atlantic states], but they are found in large flocks in all our extensive woods. Amidst the mountains and

> broken countries are great numbers of the grouse I have described; and since the settlement has been established, the quail, by following the trail of grain which is necessarily scattered through the wilderness, has migrated from the old settlements on the other side the mountains, and has become a constant resident with us. This bird was unknown here on the first peopling of the country. There is a variety of wild-fowl in every part of this State, particularly teal, and the summer [wood] duck. The latter breeds with us. Its incubation is always in temperate climates, which is the reason of its being called the summer duck.[245]

Imlay listed 111 bird species by their popular names—more than any previous writer describing the Midwest—and then gave the equivalent scientific name assigned by Linnaeus or Catesby. "In my account of the birds of this country, I shall mostly give you the Linnæan designation, in preference to Catesby's," he wrote, "though Catesby's designation is most general." Imlay's preference for Linnaeus's taxonomic system over Catesby's was a sign of the changing times.[246]

Thomas Jefferson was the first American writer to consistently employ Linnaean binomials for American birds. His 1787 *Notes on the State of Virginia* listed vernacular and Linnaean names alongside those used by Catesby and the Count de Buffon, whose thirty-six-volume *Histoire Naturelle, Générale et Particulière* . . . was the most widely cited biological compendium of the day. Jefferson also pushed the Linnaean system in his letters, and by the 1780s and 1790s American writers were citing Linnaeus not only in scientific works but also in histories, geographies, poems, and even *The Christian's Pocket Library* (New York, 1796).[247]

Jefferson was in many ways the quintessential Enlightenment figure—philosopher, scientist, statesman, and writer. When President John F. Kennedy hosted a dinner for forty-nine Nobel Prize winners on April 29, 1962, he famously called it "the most extraordinary collection of talent, of human knowledge, that has ever been gathered together at the White House, with the possible exception of when Thomas Jefferson dined alone."[248]

Jefferson's *Notes on the State of Virginia* was the first book to make Linnaean bird names well-known to American readers. Written in 1781 and reprinted more than a dozen times over the next two decades, it listed the Linnaean names of 126 American birds alongside Catesby's.

Jefferson proposed a scientific exploration of the West as early as 1783, but twenty years passed before his vision could be realized. In April 1803, while he was president, the United States secured a large portion of land from France through the Louisiana Purchase. Before the ink was dry on that document, Jefferson ordered a scientific expedition led by his private secretary, former army captain Meriwether Lewis, to explore the new territory.[249]

"The object of your mission," he wrote to Lewis on June 20, 1803, "is to explore the Missouri river, & such principal streams of it, as, by its course & communication with the waters of the Pacific Ocean, may offer the most direct & practicable water communication across this continent, for the purposes of commerce." But Jefferson also instructed Lewis that scientific "observations are to be taken with great pains and accuracy; to be entered distinctly and intelligibly for others as well as yourself." The explorers were ordered to make notes on the geography, landscape, and natural features, including "the animals of the country generally, & especially those not known in the U. S.," down to "particular birds, reptiles, or insects." Lewis selected an army acquaintance, William Clark, to lead the expedition with him and spent several weeks gathering scientific instruments and learning how to use them. The two officers met at the Falls of the Ohio to start west in mid-October 1803.[250]

Between 1803 and 1806, Lewis and Clark trekked all the way to the Pacific and back. The scope of this book, however, is concerned only with the first half of their expedition, through the spring of 1805, when they traveled from St. Louis to Ft. Mandan, North Dakota. While on the Great Plains they noted in their journals forty-four different bird species, several of which, including the trumpeter swan, least tern, and western meadowlark, they were the first to describe scientifically.

Lewis, Clark, and their men were close observers. Some of the new birds were "as carefully described as any practicing ornithologist of the day might have done," according to scientist Paul Johnsgard. Indeed, some of Lewis's bird descriptions run to six or eight hundred words and include intricate tables of measurements. He sketched several birds in his journal

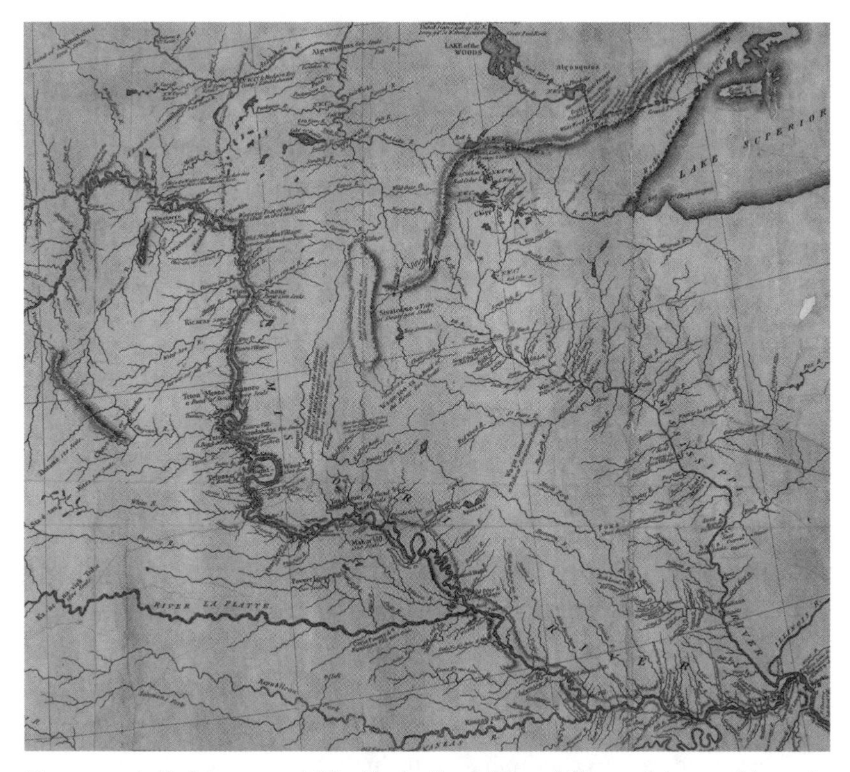

The eastern half of the map published in the first edition of *History of the expedition under the command of Captains Lewis and Clark...* by Meriwether Lewis, 1814. WHI IMAGE ID 37931

and sent live specimens back to Washington before heading further west in April 1805.[251]

Here, for example, are excerpts from Lewis's description of one species new to science, the least tern, made on August 5, 1804, in what is now Washington County, Nebraska. His idiosyncratic spelling and punctuation help us visualize Captain Lewis, tall and wiry, scratching in his notebook during a quiet moment in camp:

> I have frequently observed an acquatic bird in the cours of asscending this river but have never been able to procure one before today . . . this bird, lives on small fish, worms and bugs which it takes on the virge of the water it is seldom seen to light on trees an qu[i]te as seldom do they lite in the water and swim tho' the foot would indicate

> that they did it's being webbed . . . the weight of the male bird is one ounce and a half, it[s] [l]ength from b[e]ak to toe 7½ inches from tip to tip of wing across the back one foot seven inches and a half [the beak] is one ⅛ inch long, large where it joins the head flated on the sides and tapering to a sharp point, a little declining and curvated, a fine yellow, with a shade of black on the extremity of upper beak.

Lewis's minutely detailed physical description continues for more than three hundred words, with a table of precise measurements, before he returns to its behavior: "This bird is very noysey when flying which it does exttreemly swift the motion of the wing is much like that of kildee it has two notes one like the squaking of a small pig only on reather a high kee, and the other kit'-tee'-kit'-tee'- as near as letters can express the sound—"[252]

Careful observation and meticulous note taking didn't prevent Lewis from sometimes writing in a more conversational tone. "I saw a great number of feathers floating down the river," he noted on August 8, 1804.

> . . . They appeared in such quantities as to cover pretty generally sixty or seventy yards of the breadth of the river. for three miles after I saw those feathers continuing to run in that manner, we did not percieve from whence they came, at length we were surprised by the appearance of a flock of Pillican [American white pelican] at rest on a large sand bar . . . the number of which would if estimated appear almost incredible; they apeared to cover several acres of ground, and were no doubt engaged in procuring their ordinary food; . . . we now approached them within about three hundred yards before they {attempted to fly} flew; I then fired at random among the flock with my rifle and brought one down; the discription of this bird is as follows.

Lewis then provides more than five hundred words of painstaking description such as, "In the present subject I measured this pouch and found its contents 5 gallons of water," supplemented by two tables of measurements. His careful description of his specimen not only followed Jefferson's instructions for the expedition, but also provides a good example of fledgling American science finding its way.[253]

On September 17, 1804, Lewis and Clark recorded the first known specimen of the black-billed magpie. "One of the hunters killed a bird of the Corvus genus and order of the pica & about the size of a jack-daw with a remarkable long tale. beautifully variagated. it {has an agreeable note something like goald winged Blackbird} note is not disagreeable though loud—it is twait twait twait, twait; twait, twait twait, twait." The magpie's rarity led three members of the expedition to note it in their journals, and Lewis attempted to give it a Linnaean name (something he did infrequently). By spring, when they sent four of the live birds back to Washington, Sergeant John Ordway identified it as a magpie. One of the four survived, was displayed at Peale's Museum in Philadelphia, and became the model for Alexander Wilson's illustration of the species in his monumental nine-volume *American Ornithology,* published 1808–1814.[254]

As skies darkened and temperatures dropped, Lewis and Clark decided to winter on the Missouri opposite a village of Mandan and Hidatsa Indians. Here they met the Shoshone woman named Sacagawea ("bird-woman" in Hidatsa, the tribe among whom she was raised) who would guide them through the western mountains. Near their camp, on October 16, 1804, Lewis captured the first common poorwill known to science and was puzzled by its semidormant state:

> This day took a small bird alive of the order of the [blank space; he apparently intended to look up the Latin name later] or goat suckers. it appeared to be passing into the dormant state. on the morning of the 18th the murcury was at 30 a[bove]. the bird could scarcely move.—I run my penknife into it's body under the wing and completely distroyed it's lungs and heart—yet it lived upwards of two hours[.] this fanominon I could not account for unless it proceeded from the want of circulation of the blo[o]d.—the recarees [Arikara Indians] call this bird to'-na it's note is at-tah-to'-nah'; at-tah'to'-nah'; to-nah, a nocturnal bird, sings only in the night as does the whipperwill.

More than a century would pass before modern ornithologists confirmed the poorwill's ability to hibernate.[255]

Lewis and Clark spent a total of fourteen months on the Great Plains and captured the first detailed data on at least eight bird species unknown to Eastern naturalists: common poorwill, piping plover, western meadow-

Grouse
are
about
short
and eye.
Cock.
Cock
which
on the
and
hood
Mountains
to the Mountain
the Columbia
the Great falls
they go in large
or singular ly

the feathers about it's head
pointed and stiff some hairs
the base of the beak. feathers
fine and stiff about the ears.
This is a faint likeness of the
of the Plains or Heath
the first of those fowls
we met with was
Missouri below
in the neighbour-
of the Rocky
and from
which pass
between
and Rapids
Gorges
and
mak

hide hide remarkably close when pursued.
short flights &c.

The large Black & White Pheasant is peculiar
to that portion of the Rocky Mountains watered by
the Columbia River. at least we did not see them untill
we reached the waters of that river, nor since we have
left those mountains. they are about the size of a
well grown hen. the cantour of the bird is much
that of the redish brown Pheasant common to
our country. the tail is proportionably as long and is
composed of 18 feathers of equal length, of a uniform
dark brown tiped with black. the feathers of the
body are of a dark brown black and white. the black

Meriwether Lewis's sketch of a sage-grouse, 1806. MISSOURI HISTORY MUSEUM, ST. LOUIS

lark, greater sage-grouse, McCown's longspur, Lewis's woodpecker, trumpeter swan, and least tern. The tally for the entire expedition, including their time in the mountains and on the Pacific Coast, was twenty-five birds new to science and another twenty accurately described for the first time.[256]

Jefferson and the other Founding Fathers realized that the new nation needed an icon, a symbol to accompany its constitution and its flag, and after considering the rooster, dove, and mythical phoenix, they settled upon the bald eagle. It was not a unanimous choice. Benjamin Franklin wrote to his daughter, Sarah, on January 26, 1784:

> For my own part I wish the Bald Eagle had not been chosen the Representative of our Country. He is a Bird of bad moral Character. He does not get his Living honestly. You may have seen him perched on some dead Tree near the River, where, too lazy to fish for himself, he watches the Labour of the Fishing Hawk; and when that diligent Bird has at length taken a Fish, and is bearing it to his Nest for the Support of his Mate and young Ones, the Bald Eagle pursues him and takes it from him.
>
> With all this Injustice, he is never in good Case but like those among Men who live by Sharping & Robbing he is generally poor and often very lousy. Besides he is a rank Coward: The little King Bird not bigger than a Sparrow attacks him boldly and drives him out of the District. He is therefore by no means a proper Emblem for the brave and honest Cincinnati of America who have driven all the King birds from our Country.[257]

Franklin would have preferred to see the turkey on the Great Seal. "For the Truth [is,]" he explained, "the Turkey is in Comparison a much more respectable Bird, and withal a true original Native of America . . . He is besides, though a little vain & silly, a Bird of Courage, and would not hesitate to attack a Grenadier of the British Guards who should presume to invade his Farm Yard with a red Coat on."[258]

Franklin was not the only one who disapproved of using the bald eagle to represent the new nation. When an Ohio judge showed the Great Seal of the United States to a local Shawnee chief in 1789, he received the following reply:

> You have told me much of the peaceful intentions of the United States towards the Indians, and you show as a proof this picture. If the United States were such lovers of peace as you describe them to be, they would have chosen for their coat of arms something more appropriate and expressive of it. For example there are many good, innocent birds. There is the dove which would not do harm to the smallest creature. But what is the eagle? He is the largest of all birds and the enemy of all birds. He is proud, because he is conscious of his size and strength. On a tree, as well as in flight he shows his pride and looks down disparagingly upon all the birds. His head, his eyes,

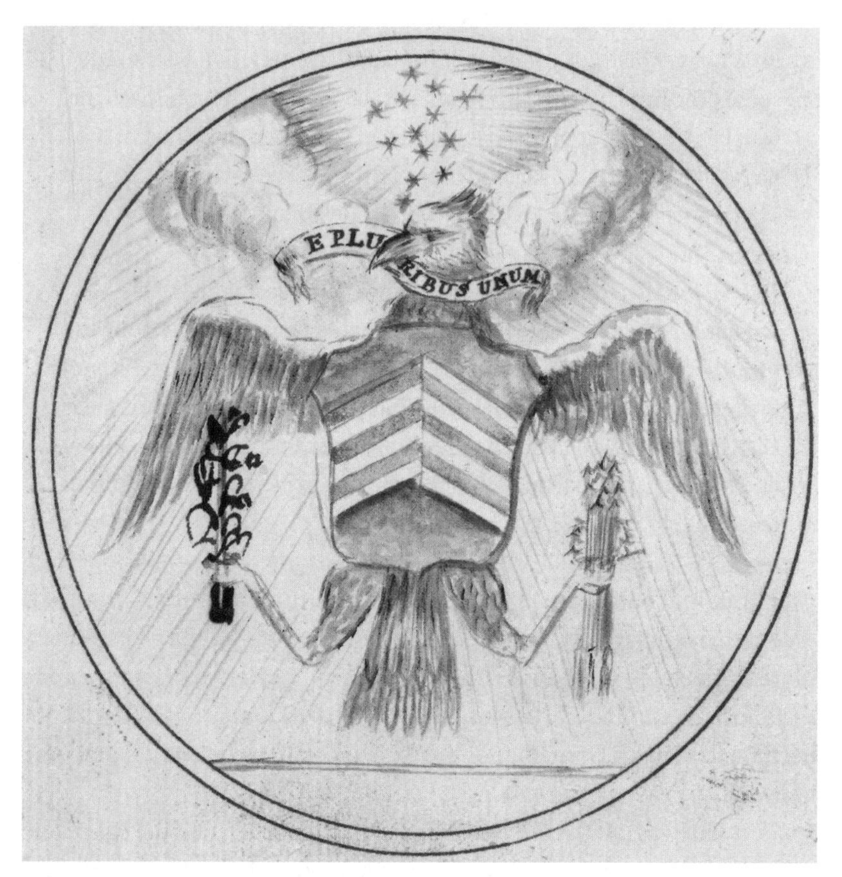

Design for the official seal of United States adopted by the Continental Congress in 1782 and used until 1842, showing the eagle, talons, and arrows that offended both a Shawnee chief and Benjamin Franklin. NATIONAL ARCHIVES IMAGES VIA WIKIMEDIA COMMONS

> his beak and his long brown talons declare his strength and hostility. Now this bird, which is terrible enough in itself you have depicted as even more dreadful and horrible. You have not only put one of the implements of war, a bundle of arrows, into one of his talons, and rods in the other, but have painted him in the most fearful manner, and in a position to attack upon his prey. Now tell me, have I not spoken the truth?[259]

The judge diplomatically replied that only enemies of the United States had to fear the eagle's ferocity, and that the new nation's friends would regard its weapons as protection.

By then, reason and science had replaced dogma and faith in the minds of most educated Americans. Jefferson's instructions and Lewis and Clark's meticulous observations reflected a worldview radically different from the one that Jesuit priests had carried into the American wilderness a century earlier.

Seventeenth-century missionaries like Louis Nicolas and Jacques Marquette saw the hand of God everywhere and, while aware of dawning scientific theories, firmly believed in a supernatural world. They also did their best to show church authorities that their writings conformed to religious dogma.

In contrast, the soldiers who followed them two generations later did their best to sound scientific, since public discourse in Philadelphia, London, and Paris had embraced naturalists such as Linnaeus. They viewed birds not as messengers between the shamanist Upper and Lower Worlds or as manifestations of God's glory but rather as objects to be studiously examined, measured, and labeled.

During the century that separated La Salle and Jefferson, the worldviews of educated observers had evolved from magic and mystery to a rational, secular perspective much closer to our own. This change in attitudes produced the first generation of scientists who could be called ornithologists.

7

Fledgling Ornithologists

Like most Americans, I knew the name Audubon before I knew anything about the man himself. In my youth, the name conjured up vague associations with birds and art—some eminent Victorian painter who'd founded a conservation group, perhaps? After college, when I landed my first job working with antiquarian books, I connected it to preposterously large and incredibly expensive works, truly big game that none of my employers ever succeeded in capturing.

When I became a rare-book librarian, I finally got to see the famous elephant folios up close and to examine them alongside their contemporaries. But not until I began research for *Taking Flight* did I actually read Audubon's books, letters, and journals. Putting this off was a great mistake, because Audubon is an enormous pleasure to read, ranking with Henry Thoreau, John Muir, and Aldo Leopold as one of America's great nature writers.

In 1807, Audubon opened a shop in William Clark's hometown of Louisville, Kentucky, shortly after the explorer returned from the West. Audubon had been born in 1785 in Haiti, the illegitimate son of a French merchant and his chambermaid. He was raised in France until 1803, when his father sent him to America to escape the draft. He settled first near Valley Forge, Pennsylvania, where he failed at farming, mining, and business while studying and painting birds.

"At a very early period of my life," Audubon recalled afterward, "I arrived in the United States of America, where, prompted by an innate desire to acquire a thorough knowledge of the birds of this happy country, I formed the resolution, immediately on my landing, to spend, if not all

John James Audubon in 1841, painted by his sons John W. and Victor Audubon. AMERICAN MUSEUM OF NATURAL HISTORY LIBRARY, IMAGE #1822

my time in that study, at least all that portion generally called leisure, and to draw each individual of its natural size and colouring."[260]

In 1807 he moved with his new bride, Lucy, to the frontier hamlet of Louisville, Kentucky, and opened a general store. Audubon lived in the Ohio Valley for the next thirteen years, spending as little time as possible behind his shop counter in order to roam the woods and wetlands collecting birds to paint. "I never for a day gave up listening to the songs of our birds," he told his children later, "or watching their peculiar habits, or delineating them in the best way that I could." He loved the wild, and claimed that "it was often necessary for me to exert my will, and compel myself to return to my fellow-beings."[261]

In 1819, Audubon, never very good at shopkeeping, finally went bankrupt. He and his wife lost all their property except his bird paintings, which the creditors considered worthless. He gave up business altogether in order to live by his wits and draw birds. A decade of wandering carried him from Louisiana to Labrador in pursuit of his art. His early paintings were lost in transit, spoiled by moisture, and eaten by rats before finally being embraced by the public in 1827. He spent the 1830s writing and publishing his famous "double elephant folio" edition of *Birds of America* and, by 1840, was comfortable and famous. In 1846, after fifty years of nearly constant travel and unremitting work, he suffered a stroke. A visiting friend reported that his mind was "all in ruins," and he died five years later at age sixty-six.[262]

Audubon lived for more than a decade in the Ohio Valley, and also made four trips through the center of the continent. In the winter of 1810–1811, he traveled down the Ohio River and up the Mississippi into Missouri; in the fall of 1820, he went down the Ohio and Mississippi into Arkansas before spending the winter in New Orleans; in September 1824 he traveled north from Pittsburgh to Lake Ontario; and from March to November 1843 he journeyed up the Missouri to the mouth of the Yellowstone, on the border of North Dakota and Montana. Everywhere he went, he painted and wrote about birds.

"I read in Audubon with a thrill of delight," his contemporary Henry Thoreau said in 1842.[263] Audubon is remembered today mostly for his vivid, life-sized bird portraits, but he was also a prolific and powerful writer. He published more than three thousand pages on American natural history, not counting his private journals and letters. English was his second language, and he never left behind French syntax or mastered the punctuation and spelling conventions of educated Americans. But his unrestrained exuberance, the contagious pleasure he conveys of being vitally awake and aware, is irresistible.

For example, here is part of his journal entry for December 10, 1820, just as he penned it while floating down the Arkansas River to rejoin the Mississippi:

> The Morning broke and with it, Mirth, all about us, the Cardinals, the Towe [towhee] Buntings, the Meadow Larks and Many Species of Sparrows, chearing the approach of a Benevolent sun shining day—. . . [Coming upon an American Indian family,] We gave them 50 cts and a Couple Loads of Gun Powder to each, brought out smiles, and a Cordial Shaking of Hands . . . Whenever I meet Indians I feel the greatness of our Creator in all its splendor, for there I see the Man Naked from his Hand and Yet free from Acquired Sorrow == . . . The

> Intrepid Hawks are extremely plenty along the Banks of the Mississipi where the feed aboundantly on the Swamp Sparrows as also on the Sturnus depradatorius [Red-winged Blackbird], some of these are so strong and daring that they Will attack some Ducks on the Wings and often carry them off several hundreds of yards to the Sand Bars—[264]

When he was writing for publication, Audubon hired a professional editor to improve his English or asked his American wife for help, and most of his journals were polished by his granddaughter before being printed. Despite their heavy-handed editorial interference, his childlike enthusiasm usually survived intact. Here, as an example, is a paragraph from his *Ornithological Biography*, the text that accompanied the famous illustrations in *Birds of America*. Audubon's initial draft has been cleaned up by his British editor, William Macgillivray, but his personality still shines through:

> Imagine yourself with a gun on your shoulder, following the windings of one of those noble streams which embellish our country and facilitate its commerce, having constantly within your view millions of birds on their way to the south, and which in the evenings fall, thick as the drops of a hail-shower, on the bordering marshes, to spend the night there in security, and by rest to restore the vigour necessary for their gaining the distant regions, whence half of them had emerged the preceding spring. Well, as you are proceeding, full of anxiety, and gazing in astonishment at the multitudes of feathered travellers, all of a sudden a larger bird attracts your eye. It sweeps along in the stillness of the autumnal evening with a rapidity seldom equalled, creating confusion, terror, and dismay along the whole shores. The flocks rise en masse with a fluttering sound which comes strangely on your ear, double and double again, turn and wind over the marsh, agitated and fearful of imminent danger. And now, closely crowded, they would fain escape, but alas! one has been singled out, and in the twinkling of an eye, the Pigeon Hawk [Merlin], darting into the middle of the flock, seizes and carries him off.[265]

Nothing like this existed in American nature writing at the time, nor would anything like it be seen again until John Muir began composing

Audubon's pigeon hawk (merlin). NATIONAL AUDUBON SOCIETY

his memoirs almost a century later. Audubon's unabashed love for "our delightful America pure from the hands of its Creator!" and his lively, intimate prose could not have been more different from the serious science just then coming into fashion. His habit of assigning human traits to birds, his direct addresses to readers ("Imagine yourself . . ."), and his frequent use of romantic imagery would never have made their way into books produced by the academic establishment of his time.[266]

That establishment was centered in Philadelphia, where most of the nation's scientific activity was concentrated in two learned societies. The first, the American Philosophical Society, was modeled on the British Royal Society, whose mission was to "encourage philosophical studies, especially those which by actual experiments attempt either to shape out a new philosophy or to perfect the old." Its members included Benjamin Franklin, Thomas Jefferson, William Bartram, and other members of the educated elite "who trafficked in learned discourse and possessed access to erudite transatlantic correspondents and membership in polite institutions."[267]

The second organization, the Academy of Natural Sciences, had been founded in 1812 by "impoverished young freethinkers, radicals and socialists" committed to the "collecting and preserving of natural curiosities." The Academy was bankrolled by businessman William Maclure and included energetic young naturalists who preferred to spend more time in the field than in the salon. In 1825, several of its most important members left Philadelphia altogether to start the utopian commune of New Harmony in Indiana, which was then considered the western frontier.[268]

Some naturalists belonged to both institutions, as well as to ones in other cities. Between 1815 and 1825, the number of American learned societies tripled, and groups in New York, New Haven, and Boston all launched scholarly journals. But Philadelphia was the undisputed center of American learning in the first half of the nineteenth century. Its scientific community closely followed the work of their European peers, sent specimens across the Atlantic, hosted Linnaeus's pupils when they visited the United States, organized expeditions to the uncharted west, created the nation's first natural history museum, and published its first scientific journal. They were forward-looking, optimistic, and patriotic and intended to demonstrate that citizens of the new republic could contribute as much to science as Europeans could.[269]

Most of these people wanted nothing to do with Audubon. When he visited Philadelphia in 1824, its leading scientists called him an "uncouth upstart" who "looked like the backwoodsmen that visit the city. His hair hung on his shoulders and his neck was open." In their eyes, he was nothing more than an eccentric frontier shopkeeper who painted birds in his spare time. Their snobbishness could be traced, in part, back to a chance encounter fourteen years earlier, when Philadelphia ornithologist Alexander Wilson visited Audubon's general store in Louisville on March 19, 1810.[270]

"One fair morning," Audubon recalled, "I was surprised by the sudden entrance into our counting-room of Mr. Alexander Wilson, the celebrated author of the 'American Ornithology,' of whose existence I had never until that moment been apprised." Wilson, like Audubon, was an impoverished immigrant who had scrambled to make ends meet while studying and drawing American birds. In 1803, he gained the support of William Bartram, the most respected natural scientist in Philadelphia, to compile a comprehensive work describing and depicting American birds. By 1805, he'd made twenty-eight drawings, and a local publisher had agreed to print his work if he could find two hundred subscribers.[271]

To recruit those, Wilson made five long journeys between Maine and Florida, including walking from Philadelphia to Niagara Falls on one trip and from Pittsburgh to New Orleans on another. His proposed book cost $120, a full year's wages for most Americans, and he found it easier to gather bird specimens than customers. His journals are peppered with acerbic comments on the ignorance and crassness of American businessmen. The first volume nevertheless appeared in 1808, and in the end he secured about 450 subscribers.[272]

Wilson was a much more careful scientist than Audubon and had the institutional backing of the nation's leading thinkers and scholars. Although Wilson's *American Ornithology* has since become overshadowed in the popular mind by Audubon's more famous book, it is a landmark in the history of American science. An obsessive perfectionist, Wilson ultimately worked himself to death seeing it through the press. He not only drew all the illustrations and wrote the accompanying texts for 278 birds, but he also oversaw the printing and hand coloring of each 11-by-15-inch plate. He finally collapsed from exhaustion in 1813 after publishing seven volumes, and his friend George Ord, vice president of the American Philosophical Society, completed the last two after Wilson's death.[273]

When Wilson stepped into Audubon's shop in the spring of 1810, he carried the first slender volumes of his book under his arm. After Audubon had admired their illustrations and confessed to his own work along similar lines, he recounted:

> Mr. Wilson asked me if I had many drawings of birds. I rose, took down a large portfolio, laid it on the table, and shewed him, as I

Wilson's pigeon hawk (merlin). FROM *AMERICAN ORNITHOLOGY* BY ALEXANDER WILSON

would shew you, kind reader, or any other person fond of such subjects, the whole of the contents, with the same patience with which he had shewn me his own engravings. His surprise appeared great, as he told me he never had the most distant idea that any other individual than himself had been engaged in forming such a collection.

Audubon claimed that while Wilson was in Louisville: "[I] exerted myself as much as was in my power, to procure for him the specimens which he wanted. We hunted together, and obtained birds which he had never before seen; but, reader, I did not subscribe to his work, for, even at that time, my collection was greater than his." Audubon was probably also too poor to afford a copy.[274]

Wilson left Louisville in a sour mood. He complained in his diary, "Everyone is so intent on making money, that they can talk of nothing else; and they absolutely devour their meals, that they may return sooner to their business. Their manners correspond with their features." And he was irritated by Audubon, whom he didn't consider a serious naturalist. On March 19, Wilson had confided to his diary, Audubon's "drawings in crayons—very good. Saw two new birds he had," but when he left town four days later he wrote that he had "neither received one act of civility from those to whom I was recommended, one subscriber, *nor one new bird*; though I delivered my letters, ransacked the woods repeatedly, and visited all the characters likely to subscribe. *Science or literature has not one friend in this place*." He went back to Philadelphia with a bitter taste in his mouth, and probably disparaged Audubon to his colleagues.[275]

Audubon continued to paint birds for more than twenty-five years before publishing his own work. Because he drew every bird at life size, including the turkey and great blue heron, he had to go to England to find a publisher who could print large enough pages. It took twelve years to complete the project, from 1827 to 1838. Each page measured thirty by forty inches and was printed from a copper engraving before being colored by hand.

Like Wilson, Audubon oversaw every step in the process. Plates were issued to subscribers in groups of five, with twenty five-plate numbers being bound together into a volume. With a total of 435 plates measuring the size of a kitchen table, the book filled four massive volumes, each of which weighed forty pounds. Once the plates were printed, Audubon wrote nine prose volumes of *Ornithological Biography, or An Account of the Habits of the Birds of the United States of America,* which filled more than three thousand pages. When it was finished in 1838, *Birds of America* was the largest book ever created.

It was also enormously expensive. At $1,000 per copy, the book cost more than the annual budgets of some of the libraries to which he pitched it. Audubon sold fewer than two hundred copies, most of them to English aristocrats and American millionaires. In 1840, he published a cheaper edition in seven octavo volumes that integrated the plates and the text, but even these were far beyond the reach of most Americans' purses. In recent years, complete sets of the double elephant folio have fetched several million dollars at auction and the 1840 "cheap" edition nearly $100,000.[276]

As the individual parts appeared during the 1830s, reviewers everywhere preferred Audubon's work to Wilson's because it was larger, more beautiful, and contained almost twice as many birds. The critics' praise for it sparked the Philadelphia establishment into a defense of Wilson. "As soon as his popularity and success began to check the sales of Wilson's work," a friend of Audubon recalled, "Ord and a few others, aided by interested publishers, began a systematic series of attacks." They criticized Audubon in letters and articles, accused him of plagiarism, and derided his lack of credentials. Wilson's friend George Ord blackballed Audubon from the American Philosophical Society just when becoming a member would have helped him enroll more subscribers. The partisan bickering between the two camps continued for decades after both artists had died.[277]

Wilson and Audubon's encounter in Louisville in March 1810 was more than a clash of egos. It was a clash of cultures, of insiders versus outsiders. On one side were curious laypeople such as frontier shopkeepers and farmers who reacted emotionally, spiritually, and creatively to the birds that surrounded them. They wanted not just facts, but also to know what the facts meant and how they could be applied. On the other side were genteel, urban scientists who focused on gathering, labeling, and arranging data. Their main concerns were making sense of the deluge of new information about birds coming in from explorers like Lewis and Clark and proving that American scientists were as talented as European ones.[278]

Between 1800 and 1860, federal and state governments sent eighty-two expeditions across the continent to collect scientific data. The Philadelphia institutions also financed exploration of their own. The young naturalists who went on these dangerous journeys published more than 150 reports and articles about birds and brought more than twenty-nine thousand bird skins back to Philadelphia's museums.[279]

As these investigators returned from their travels around the new nation, the Philadelphia naturalists compared the dead specimens with those already in their museums, hypothesized explanations for differences, published theories in their journals, and debated conclusions at their meetings. They also created the country's first stable vocabulary of bird names and classified them into a rational taxonomy based on Linnaeus. American ornithology—as distinct from casual nature writing about birds—was born. As ornithologist Tod Highsmith points out, wholesale hunting by museums was driven not by greed or bloodlust, but by the need to gather enough physical evidence for scientists to delineate species, split them apart or lump them together, and forge a lasting, reliable taxonomy of American birds. It was an effort that continued for more than a century.[280]

Audubon was left out of these developments for most of his career. As an artist and writer, he wanted to depict birds in action and convey their vitality, yet at the same time he felt pressure to meet the new scientific standards. "I knew well," he wrote, "that closet naturalists would expect drawings exhibiting, in the old way all those parts that are called by them necessary characteristics; and to content these gentlemen I have put in all my representations of groups always either parts or entire specimens, showing fully all that may be defined of those particulars."[281]

But Audubon couldn't resist sprinkling his texts with asides to the reader, expostulations of amazement, colorful tales, and other literary devices that the scientists deplored. He even interspersed among his bird descriptions thirty anecdotes about his own personal adventures. This alienated him from the emerging scientific community, who were more interested in precise measurements of bills and feathers, or dry taxonomical arguments about Latin binomials, than in joyful outbursts over nature's wonders.

Audubon's writings and paintings may give us more pleasure, but the Philadelphia naturalists gave us our modern understanding of birds. Willing to risk any hardship to collect new facts, they trekked tens of thousands of miles on foot, in canoes, by steamboat, and on horseback, carrying notebooks and specimen cases across the Plains, over the Rockies, and down the Pacific Coast. Their letters, diaries, memoirs, and official reports planted the seeds for how we understand birds today. Every modern birder memorizes certain things about birds and overlooks others, calls a bird by one name instead of another, because naturalists thought in specific ways in Philadelphia two hundred years ago.

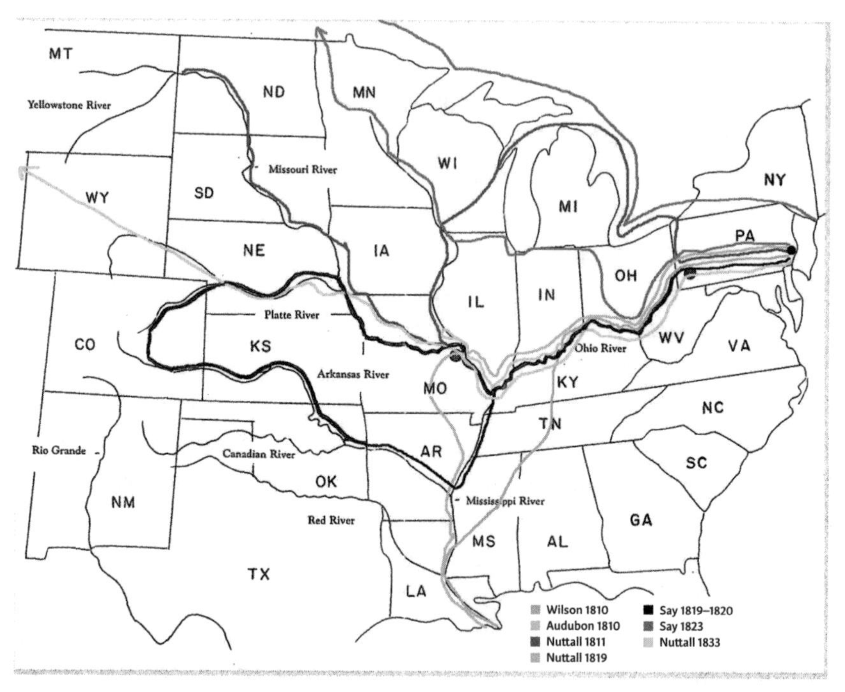

Approximate routes of naturalists discussed in this chapter, 1810-1833. MICHAEL EDMONDS

This generation of early ornithologists is perhaps best typified by Thomas Say (1787–1834) and Charles-Lucien Bonaparte (1803–1857), a pair of friends who extended and improved Wilson's book on American birds. Say, a native Philadelphian who crisscrossed the Midwest in pursuit of birds, was tall, thin, easygoing, and eccentric. His friend Bonaparte, a French aristocrat, was short, round, arrogant, and conventional. But they were each radical in their politics and, more importantly, obsessive about scientific research.

Say came from a well-known Quaker family with connections to John and William Bartram, in whose famous gardens he spent many childhood hours. Like Audubon, Say failed at business as a young man because he spent too many hours outdoors instead of in his shop. He abstained from smoking and drinking, lived stoically, and was so likable that friends claimed, "It was impossible to quarrel with him." In 1812 he helped found the Academy of Natural Sciences and edited its journal for many years.[282]

But while Say was very affable, he was also an odd duck. For long periods, he spent six cents a day on food and slept on the floor of the academy's museum, under its exhibit cases. "To him," George Ord recalled, "the season of midnight was the hour of prime, it was the time of stillness and tranquility; and so greatly did he enjoy these vigils, that he not unfrequently prolonged them, even during the summer, until the approach of day." He was one of the Philadelphia naturalists who turned their backs on eastern culture in 1825 to start a socialist commune on the Ohio frontier.[283]

Bonaparte, in contrast, was a nephew of the French emperor Napoleon, who exiled his relatives for their radical politics. After living for a while in Italy, the family bought 1,700 acres in New Jersey and created a luxurious estate, where the young nobleman developed a love for nature. As a teenager, he assembled an enormous private collection of birds, mammals, and insects, as well as a library of the latest and most expensive European scientific books. Though raised in America, Bonaparte always felt more comfortable speaking French or Italian than English, and he traveled often in Europe among his aristocratic peers. But his wealthy connections didn't prevent him from being a passionate democrat and supporting revolutions wherever the common people rose up against tyranny.

Bonaparte moved to Philadelphia in 1824, at the age of twenty-one, and was immediately accepted into the community of learned naturalists. His primary interest was classification: how all his specimens, and by extension all of nature, should be properly arranged. He had never spent time in the wilderness and didn't intend to. He hired other people to be his eyes and ears.[284]

When the two men met, Say had just returned from major expeditions across the heart of the continent. In 1819 and 1820, he'd served as chief zoologist on Stephen Long's eighteen-month journey through the southern plains to Colorado, the first government expedition to include professional scientists. His detailed accounts of birds from that expedition fill technical footnotes throughout its two-volume report.

Following that journey, Say was chosen again by Long to serve as scientist on an 1823 expedition west through Minnesota and across the northern plains to Lake of the Woods. The official report on that trip includes his 150-page appendix on animals and plants (but no birds). By the time he met Bonaparte, Say had journeyed more than 2,500 miles through the Midwest and Great Plains and was one of the most widely traveled naturalists in America.[285]

Charles-Lucien Bonaparte in 1849, from a lithograph by Thomas Herbert Maguire. WIKIMEDIA COMMONS

Bonaparte had just decided to write a book about the western birds missing from Wilson's text, and Say had conveniently brought back the skins of precisely those birds. He wasn't primarily interested in birds—insects were Say's passion—and his collection of skins hadn't been shared with anyone. It was just what Bonaparte needed.

He also needed an artist to illustrate the book. Luckily, Say's assistant on the 1819–1820 expedition to the Rockies, Titian Ramsay Peale (1799–1885), was available. He was the youngest son of painter Charles Willson Peale, had been elected to the Academy of Natural Sciences at age eighteen, and combined the expertise of a naturalist with the skills of an artist. In 1824, the trio began work on Bonaparte's *American Ornithology, or The Natural History of Birds Inhabiting the United States, Not Given by Wilson . . .*[286]

At this date it's difficult to untangle how much of the initial work was Bonaparte's and how much was Say's. For the most part, the ambitious aristocrat claimed authorship, which was fine with the self-effacing Quaker. However, Bonaparte also credited his partner in passages such as this: "The bird now before us was brought from the Arkansaw river, in the neighbourhood of the Rocky Mountains, by Major Long's exploring party, and was described by Say under the name of *Troglodytes obsolete.*" Before the first volume appeared in 1825, Say read and rewrote nearly every page of Bonaparte's awkward English. After Say moved to New Harmony, Indiana, in the fall of 1825, Bonaparte persuaded another friend, John Godman, to clean up his prose.[287]

All three men—Say, Bonaparte, and Peale—were thorough scientists. In the words of a colleague, Say "was an advocate for that doctrine which attached exclusive importance to the evidence of the senses; fact alone was the object which he thought worthy of his researches." It could have been said about nearly all the Philadelphia naturalists.[288]

Say's Flycatcher (lower right), named for Thomas Say and drawn by Titian Ramsey Peale for Bonaparte's *American Ornithology, or, The Natural History of Birds Inhabiting the United States, Not Given by Wilson*. UNIVERSITY OF WISCONSIN DIGITAL COLLECTIONS

Bonaparte devoted his own energy to meticulously describing each bird's anatomy, naming it accurately, and classifying it within "several genera, families, and especially orders, that will admit of a disposition in a regular series, and yet remain in strict conformity to nature." He was especially concerned about regularizing birds' names, to rectify what he called "a Tower of Babel." In an article he wrote on nomenclature (a term that he coined), Bonaparte complained: "If authors are to be permitted to change specific names under the excuse of improving them, there will be no end to their alterations, and our systems will be involved in utter chaos."[289]

In 1828, Bonaparte decided to return to Europe. He saw volumes two and three of his *American Ornithology* through the press that year, and in addition published a lengthy treatise on the classification of birds. He engaged William Cooper of New York, a cousin of the novelist James Fennimore Cooper, to oversee the editing and printing of the fourth volume in his absence. That work took four years, since the book's author was in Europe, its editor in New York, and its illustrator and printer in Philadelphia. Meanwhile, new birds were still being discovered, and Bonaparte would have written a fifth volume except that, after eight years of frustrating labor for minimal returns, his publisher backed out of the project.[290]

Bonaparte's four volumes were 20 percent larger and even more luxurious than Wilson's and included sixty additional species (Wilson had depicted 278). While Bonaparte's books were being published in 1828 and 1829, George Ord issued a new edition of Wilson in three smaller volumes with a large atlas of plates; later publishers often combined the two works into a single set of "Wilson and Bonaparte's birds." Before leaving America in 1829, Bonaparte also issued a 300-page *Genera of North American Birds, and a Synopsis of the Species Found within the Territory of the United States* . . . that arranged the 382 known species into new groupings. It was the pinnacle of ornithological classification up to that time.[291]

Another excellent example of Philadelphia's state of mind is Bonaparte's and Say's friend Thomas Nuttall (1786–1859). He traveled even farther than Say, through much of the Midwest, and had even more impact than Bonaparte on what most Americans knew about birds.

Nuttall grew up in rural Yorkshire, England, and apprenticed with a Liverpool printer. Setting type for natural history books ignited in him a passion for botany, and just after his twentieth birthday, he sailed for America to explore its plants. Nuttall began botanizing as soon as he stepped on shore and for most of the next thirty-five years worked his way ever farther west, until he'd reached Hawaii in pursuit of rare plants.

In 1810, American Philosophical Society vice president Benjamin Smith Barton hired Nuttall to travel to St. Louis and then up the Missouri through Kansas, Nebraska, and South Dakota to the mouth of the Yellowstone in pursuit of specimens. It was an activity "to which he appears singularly devoted," another member of the expedition wrote,

> and which seems to engross every thought, to the total disregard of his own personal safety, and sometimes to the inconvenience of the party he accompanies. To the ignorant Canadian boat-men, who are unable to appreciate the science, he affords a subject of merriment; le fou [the Fool] is the name by which he is commonly known. When the boat touches the shore, he leaps out, and no sooner is his attention arrested by a plant or flower, then every thing else is forgotten. The inquiry is made "ou est le fou?" "where is the fool?" "il est apres ramassee des racines," "he is gathering roots."

At one point, his colleague continued, Nuttall became lost in the wilderness and "unable to go a step farther, he laid himself down with resignation and would inevitably have died had he not been found by a friendly Indian who placed him in his canoe and rowed him down the Missouri River to the first settlements of the white men." Novelist James Fennimore Cooper parodied him in his novel *The Prairie* as Obed Battius, the first absentminded professor in American literature.[292]

Nuttall journeyed through the Midwest three more times, making notes on plants and birds everywhere he went. In 1816 and 1817, he floated down the Ohio and swung through the Southeast botanizing; on his return, he was elected to membership in both the American Philosophical Society and the Academy of Natural Sciences. In 1818 and 1819, he again went down the Ohio, this time crossing the Mississippi into Arkansas on "an arduous and perilous journey of more than five thousand miles mainly over a country never visited before by scientific explorers and still in the undisputed possession of the wild Indian." Finally, in 1833 Nuttall resigned a lucrative teaching job at Harvard and signed up for a three-year expedition down the Ohio, up the Kansas River, across the Plains, over the Oregon Trail to Seattle, and then by sea to Honolulu and southern California.[293]

Nuttall shared his Philadelphia patrons' mission to collect data accurately, name things unambiguously, and classify them systematically. "It is only by attentive, careful, and diversified observations," he wrote in 1831, "that we can ever expect to arrive at any certain knowledge concerning the animals which live around us."[294]

When he wrote those words, Nuttall was at Harvard completing the first volume of a popular book on American birds, his *Manual of the Ornithology of the United States and of Canada*. His goal was to produce a "compendious and scientific treatise on the subject, at a price so reasonable as to permit it to find a place in the hands of general readers." He intended to democratize the work of "the immortal Wilson and of the justly celebrated Audubon," whose large folios were rare, expensive, and inaccessible to all but the rich. Even their cheaper octavo editions, running to several volumes with hand-colored plates, were beyond the reach of most Americans. So Nuttall produced two squat, seven-by-four-inch volumes (one for land birds, the other for waterbirds) printed on inexpensive paper and illustrated with simple line drawings.[295]

His text was informal enough for general readers but technical enough for scientific ones. Lists, names, and hard facts abounded, but they were

Audubon's illustration of the Yellow-billed Magpie (center), named *Corvus nuttalli* by Audubon in honor of Thomas Nuttall (now called *Pica nuttalli*). NATIONAL AUDUBON SOCIETY

interspersed with personal stories from his travels. In addition to quoting earlier authorities and listing data, Nuttall also wrote about his own encounters with birds while hiking and paddling thousands of miles through the interior. His life was so full of birds, in fact, that his progress on the manuscript was sometimes delayed by a pet brown thrasher he kept in his Cambridge study that had "a playful turn for mischief and interruption, in which he would sometimes snatch off the paper on which [Nuttall] was writing."[296]

The leader of the Philadelphia naturalists, George Ord, probably had Nuttall in mind when he wrote,

> There are two classes of readers to whom the major part of writers on zoology, of the present day, address themselves, the reader for pastime, and the scientific naturalist. Now, that it is possible to conciliate the good opinion of both these classes is proved by the success of some publications of recent date, in which strict attention has been paid to nomenclature, arrangement and definition, and in which the habits of the animals have been detailed with all the fidelity of truth, and in all the charms of diction.

In 1841, after thirty-five years in America, Nuttall returned to England when an uncle left him a modest estate. He built a conservatory, imported and raised plants, and retired from public life. When he died there in 1859, his name had been immortalized in three North American birds: Nuttall's woodpecker (*Picoides nuttallii*), the common poorwill (*Phalaenoptilus nuttallii*), and the yellow-billed magpie (*Pica nuttalli*). The last was bestowed by Audubon, who opened his description of it with this praise: "I have conferred on this beautiful bird the name of a most zealous, learned, and enterprising naturalist, my friend Thomas Nuttall, Esq., to whom the scientific world is deeply indebted for the many additions to our zoological and botanical knowledge which have resulted from his labours. It is to him alone that we owe all that is known respecting the present species, which has not hitherto been portrayed."[297]

Naming species after colleagues was just one aspect of early-nineteenth-century science in America. The community of people who had the means to conduct explorations, build collections, analyze data, publish findings, and debate controversies was fairly small, and the major players all knew one another. This network connecting birds in the wild, naturalists in cities, and scientific knowledge in print can be seen in the history of the evening grosbeak.

On the evening of April 7, 1823, an Ojibwe boy downriver from Sault Ste. Marie, Michigan, heard a strange cry in the bush. Moving soundlessly, he discovered a flock of small gray-and-yellow birds unlike any he had ever seen before. He shot one with his bow and arrow and ran to his family's maple-sugaring camp, where he learned it was called *paushkundamo*, meaning "berrybreaker."

Knowing that the US Indian agent at the Sault was interested in natural history, the Ojibwe family brought the dead *paushkundamo* to the white man's fort. Taking the bird in hand, Henry Rowe Schoolcraft realized that it was a species unknown to Wilson or the other ornithologists whose works he had studied.

Schoolcraft (1793–1864) had been raised near Albany, New York, where he became interested at an early age in chemistry and geology. During the winter of 1818–1819, he had explored the lead mining region of Missouri and Arkansas and published an account of its geology. In 1820, he served as scientist on an expedition from Detroit around Lake Huron and Lake Superior to the sources of the Mississippi River in Minnesota. The next year, he traveled by canoe from Detroit down the Maumee and Wabash Rivers to the Ohio and then crossed southern Illinois on horseback to attend the 1821 Chicago treaty councils with the Ottawa, Ojibwe, and Potawatomi nations. In 1822, he was appointed Indian agent at Sault Ste. Marie, where he married and lived among the Ojibwe for the next eleven years. His journals throughout his travels and residence in the Midwest are laced with close observations of birds.[298]

On that spring evening in 1823, Schoolcraft skinned the grosbeak and at the first opportunity sent it to Bonaparte's editor, William Cooper, at the New York Lyceum of Natural History. Cooper, too, saw that it was an unnamed species and conferred upon it the Latin binomial *Fringilla vespertina* ("twilight finch") and the colloquial English name, evening grosbeak. After measuring it and recording its technical data, he filed the bird skin among the Lyceum's collections. In 1825, he described it to the world in a short article, "Description of a New Species of Grosbeak, Inhabiting the Northwestern Territory of the United States," published in the Lyceum's *Annals*.[299]

The process by which the grosbeak was "discovered" reflects the goals of science summarized in a popular textbook of the day: "To distinguish and describe the objects of nature, to examine their appearance, structure, properties, and uses, and to collect, preserve, and arrange them."[300] These activities privileged certain mental habits—examining, labeling, and cataloguing—while precluding others. Nuttall, Say, and Schoolcraft did not pray to the birds for guidance, encounter them on vision quests, or even see the hand of God in them.

For example, when residents near Albany were frightened by an unusually large flight of Carolina parakeets one winter, Philadelphia professor Benjamin Smith Barton wrote, "The more ignorant Dutch settlers were exceedingly alarmed. They imagined, in dreadful consternation, that it portended nothing less calamitous than the destruction of the world." Between the lines, one can almost hear the professor's Philadelphia readers snickering.

But at the time, most Americans knew nothing about the science of ornithology. Many believed that swallows spent the winter hibernating in mud beneath frozen lakes and streams, like frogs. Reputable eighteenth-century witnesses even claimed to have seen them enter or emerge from the water when spring arrived. The idea was so firmly entrenched in the popular mind that Philadelphia physician William Caldwell weighted two swallows and sunk them in the Schuylkill River for three hours to disprove it once and for all. "Our birds," he reported, "were reduced not to a state of torpidity, or suspended animation, but of absolute death." The idea of hibernating swallows was held up by naturalists as a symbol of popular ignorance, and it became "as much as a virtuous ornithologist's name is worth to whisper hibernation, torpidity, and mud."[301]

But ordinary Americans didn't grasp the value that naturalists set on scientific research. For example, Nuttall's shipmates couldn't understand why he had left the comforts of civilization to collect plants and shells at the ends of the earth. A veteran sailor explained to his puzzled comrades how science worked:

> "I've seen them colleges, and know the ropes," the old sailor began. "They keep all such things for cur'osities, and study 'em, and have men a' purpose to go and get 'em. This old chap knows what he about. He a'n't the child you take him for. He'll carry all these things to the college, and if they are better than any that they have had before, he'll be head of the college. Then, by-and-by, somebody else will go after some more, and if they beat him, he'll have to go again, or else give up his berth. That's the way they do it.[302]

Understanding birds primarily as scientific objects led academics to act in specific ways toward them, especially to shoot and dissect them. Pull out a gun today on an Audubon Society field trip and you'd probably be beaten with field guides and tied up with binocular straps. But every nineteenth-century bird-watcher believed that "the double-barreled shotgun [was their] main reliance" and that their goal was "the destruction, as a rule, of small birds, at moderate range, with the least possible injury to their plumage."[303]

That was the advice of Elliott Coues, dean of late-Victorian ornithologists. Besides having studied and collected birds from Labrador to California between 1863 and 1881, Coues was instrumental in systematizing the scientific data about them. His *Key to North American Birds* was the bible of both professional and amateur ornithologists for thirty years after its first appearance in 1872. In 1883, he cofounded the American Ornithologists' Union (AOU), the first nationwide professional association for ornithology, and imposed his will on its all-important committee on classification and nomenclature. For an entire generation, everyone who spoke or wrote about birds came under his influence.[304]

Although Coues was a generation younger than Audubon, Say, and Nuttall, his guide, *Field Ornithology*, describes the practices that they, too, employed in the wild. His book was full of blunt advice on firearms, hunting, and specimen preparation that those of us who love the birds at our backyard feeders would find revolting. It should be remembered, though, that Coues provided advice on how to kill birds because gathering specimens helped scientists make sense of the world. That didn't mean he was writing a dry technical treatise; on the contrary, he begged "the privilege of waiving formality, that he may be allowed to address the reader very familiarly, much as if chatting with a friend on a subject of mutual interest," albeit in the Victorian third-person voice.[305]

Coues recommended loading one's shotgun with the smallest ammunition possible: "When unnecessarily large, two evils result: the number of pellets in a load is decreased, the chances of killing being correspondingly lessened; and the plumage is unnecessarily injured, either by direct mutilation, or by subsequent bleeding through large holes." He preferred mustard-seed, also called dust shot, as he explained:

a small bird that would have been torn to pieces by a few large pellets, may be riddled with mustard-seed and yet be preservable; moreover, there is, as a rule, little or no bleeding from these minute holes, which close up by the elasticity of the tissues involved. It is astonishing what large birds may be brought down with the tiny pellets. I have killed hawks with such shot, knocked over a wood ibis at forty yards and once shot a wolf dead with No. 10, though I am bound to say the animal was within a few feet of me.[306]

Sentimental birders will shudder at Coues's suggestion for handling birds that were not killed by the first blast. "Squeeze the bird tightly across the chest," he advised,

under the wings, thumb on one side, middle finger on the other, forefinger pressed in the hollow at the root of the neck, between the forks of the merrythought [wishbone]. Press firmly, hard enough to fix the chest immovably and compress the lungs, but not to break in the ribs. The bird will make vigorous but ineffectual efforts to breathe, when the muscles will contract spasmodically; but in a moment more, the system relaxes with a painful shiver, light fades from the eyes, and the lids close.

He admits that "it will make you wince the first few times" and warns that it will not work with large birds: "I would as soon attempt to throttle a dog as a loon, for instance, upon which all the pressure you can give makes no sensible impression." He suggests, instead, that you should "settle the matter . . . by piercing the brain with a knife introduced into the mouth and drawn upward and obliquely backward from the palate."[307]

Having finally killed a bird in the field, naturalists stuffed the bullet holes with cotton and rolled a sheet of stiff paper into a funnel shape. Then, Coues instructs, "Setting the wings closely, adjusting disturbed feathers, and seeing that the bill points straight forward, thrust the bird head first into one of these paper cones, till it will go no further, being bound by the bulge of the breast. Let the cone be large enough for the open end to fold over or pinch together entirely beyond the tail."[308]

Once home, the collector transformed the dead creature into a museum specimen: "Lay the bird on its back, the bill pointing to your right elbow. Take the scalpel like a pen, with edge of blade uppermost, and run a straight furrow through the feathers along the middle line of the belly, from end of the breast bone to the anus." The rest of this process, which ends by peeling the bird's skin back over its skull, is perhaps best left to the imagination rather than quoted in all its gory detail.

The interior of the bird skin was then brushed with arsenic to preserve it from insects and rot. "Use dry powdered arsenic," Coues recommended, "plenty of it, and nothing else. There is no substitute for arsenic worthy of the name, and no preparation of arsenic so good as the simple substance." Although a single gram of arsenic can be fatal to humans, he minimized the danger of this poison, if used with care: "I never feel better than when working daily with arsenic."[309]

Coues suggested that killing and skinning a dozen birds was an average day's work for a naturalist in the field, and that many years could be profitably spent building one's collection this way. "How many birds of the same kind do you want?" he asked, rhetorically. "*All you can get*—with some reasonable limitations; say fifty or a hundred of any but the most abundant and widely diffused species." This was because "birdskins are a medium of exchange among ornithologists the world over; they represent value—money value and scientific value. If you have more of one kind than you can use, exchange with someone for species you lack; both parties to the transaction are equally benefited."[310]

It's important to remember that Coues was not being a cold-hearted brute. He also told readers to "let all your justifiable destruction of birds be tempered with mercy; your humanity will be continually shocked with the havoc you work, and should never permit you to take life wantonly . . . Bird-life is too beautiful a thing to destroy to no purpose: too sacred a thing, like all life, to be sacrificed, unless the tribute is hallowed by worthiness of motive. 'Not a sparrow falleth to the ground without His notice.'" But science could not advance without widespread collecting of specimens to be analyzed.[311]

Thure Kumlien of Jefferson County, Wisconsin, was one of the collectors for whom Coues was writing. Kumlien had emigrated to the American heartland in 1843 after studying science at the University of Uppsala in Sweden, where Linneaus had founded modern taxonomy. Kumlien loved natural history from boyhood, and as an adult he selected his Wisconsin farm in part because a map showed that it promised excellent birding and botanizing. He and his young wife walked seventy miles overland from Milwaukee and erected a log cabin miles from the nearest neighbor, beside a patch of oak woods above Lake Koshkonong. "It was one of the nicest log cabins around here," he recalled, "for we had an extra bedroom besides our one big room. Under the stairs we had a pantry, which was more than most of the pioneers had." They lived in it long after more comfortable wood frame houses became available.[312]

"When I came here," Kumlien told a friend, "I did not understand anything about farming, or how to handle an axe, or a plow." He bought eighty acres anyway and was soon turning them with a pair of oxen. He "preferred oxen to horses, for if a rare bird came with his vision, he could leave them in the furrow and hunt the bird." He spent his free time collecting birds, sending specimens back to his professor in Sweden, and selling skins to Eastern collectors to augment his income. His 1847 journal reveals the conflicting claims on his time:

> May 27, Planted 7 small pails of potatoes. Fixed four bird skins for an Englishman.
>
> July 4, America's high festival day. In afternoon to store with five pounds butter. Shot one duck . . .
>
> Nov. 6, Cleaned manure out of the stable. Rain and cold with thick weather and strong storm. Wrote and sent Friday (yesterday)'s letter to J.G. Bell, New York, about birds.
>
> Dec. 11, Butchered a pig in forenoon, went after an eagle in the afternoon.

The next spring (1848), Kumlien acquired a reprint of Alexander Wilson's *American Ornithology*, his first book about North American birds. He also wanted but couldn't afford the cheap edition of Audubon, which cost $30.[313]

In 1851, Kumlien reached out to Dr. Thomas M. Brewer of the Smithsonian Institution, who connected him with other ornithologists. Their correspondence fulfilled his need for scientific colleagues—"I have not seen anybody that takes any interest in anything else but wheat, potatoes and corn," he confessed—and lasted nearly three decades. Brewer connected him with ornithologists in America, and from his log cabin on the Wisconsin prairie, Kumlien carried on a multilingual correspondence with scientists in Stockholm, Berlin, London, Milan, and Leyden, as well as Washington, Philadelphia, New York, Boston, and Cambridge. In a typical transaction in 1874, he supplied fifty nests, eggs, and skins to Professor Louis Agassiz at Harvard's Museum of Comparative Zoology. He also joined several scientific societies, but his extreme shyness, lack of funds, and farm chores kept him from traveling to their meetings.[314]

Kumlien never learned to be a successful farmer, and supplying bird specimens to cultural institutions produced welcome income. "I am glad to get fifty cents apiece for yellow-headed blackbirds skins," he wrote to Brewer, "and I wish I could sell many for that price. It is easier for me to kill and skin a bird than it is to go out and work hard for fifty cents a day for a farmer." In 1870, the state hired Kumlien to create collections for its university in Madison and the Wisconsin "normal schools," predecessors of the University of Wisconsin system. The campus in Platteville, Wisconsin, for example, received more than two hundred professionally mounted and labeled bird skins to be used in teaching zoology.[315]

In 1881, as Kumlien approached old age, the Wisconsin Natural History Society hired him as a curator for its museum (which became the Milwaukee Public Museum in 1883). His main duties were to classify, arrange, stuff, mount, label, and exhibit the Society's collections of birds and mammals, and to answer questions from the public. In 1885, as many as four thousand visitors a day viewed his exhibit. Kumlien's contract also specified six to eight weeks off each spring to return to Lake Koshkonong and collect specimens for the museum. One of his protégés there was taxidermist Carl Akeley, whose later exhibits at the Field Museum and New York's Museum of Natural History educated millions.[316]

Kumlien was killed by the work that he loved. He was unpacking a shipment of preserved skins from South America when he apparently inhaled the preservative with which they'd been treated, a chemical com-

Thure Kumlien at his work table at the Milwaukee Public Museum in 1887. WHI IMAGE ID 69876

pound of mercury and chlorine. Although he'd worked with arsenic for decades and knew how to protect himself, regular exposure to it may have weakened his system sufficiently for this unfamiliar toxic compound to produce fatal results.[317]

Every Midwestern state had a pioneer bird expert like Wisconsin's Thure Kumlien, though few of them possessed his early scientific training or his international connections. But documenting the natural history of one's new home felt almost like a civic or patriotic duty to many frontier settlers, as well as a pleasure. They spent most of each day outdoors in the company of birds on their farms and wanted to learn and share as much as possible about them. The monumental scholarly works on American birds, such as Brewer, Baird, and Ridgeway's *A History of North American Birds* (four editions, 1874–1905) and Arthur Cleveland Bent's *Life Histories of*

North American Birds (twenty-one volumes, 1919–1968), depended heavily on contributions from local collectors such as Thure Kumlien.

Bird collecting became a popular hobby during the second half of the nineteenth century: twenty-three of the original members of the American Ornithologists' Union owned personal collections numbering between 25,000 and 60,000 dead birds. Louis Bishop of Pasadena, California, amassed 49,000 bird skins; J. H. Fleming of Toronto built a special fireproof building for his collection of 33,000 specimens; H. B. Conover of Chicago owned 10,500 skins, mostly waterfowl and game birds. Those numbers sound large, but since millions of ordinary Americans also enjoyed hunting and collecting birds, they are actually a tiny fraction of all the birds shot, turned inside out, rubbed with arsenic, and stuffed according to Coues's instructions.

Museum collectors were in the field with their shotguns alongside the private hunters. By 1860, the Academy of Natural Sciences in Philadelphia, the Boston Society of Natural History, Harvard College, the Smithsonian, and the University of Michigan each possessed museum collections totaling tens of thousands of bird skins. By 1883, there were seventy-three museums in the United States with collections of birds, and four full-time, salaried bird curators. The American Museum of Natural History in New York led the pack, with 685,000 dead birds. The Smithsonian and Harvard each had a quarter of a million. Besides forming expeditions to kill and collect more, the museums solicited donations from individuals. Between 1883 and 1933, an AOU historian concluded, "We see the small private collections in one way or another falling into the hands of the larger collectors, and then the collections of the latter becoming the property of museums by gift, purchase, or bequest." By 1933, the twenty-five largest American museums owned a total of 1.8 million dead birds.[318]

Despite those seemingly huge numbers, the greatest threats to bird populations in the late nineteenth century didn't come from private citizens amassing collections or from scientists harvesting specimens for museums, but rather from market hunting and habitat loss (see chapter 10).

With so much research data available in museum cabinets and drawers, ornithologists published more than twenty thousand journal articles and monographs over the next fifty years. The institutional collections finally enabled the "closet naturalists" belittled by Audubon to figure out which birds were separate species, what they should be called, and how they should be classified. Coues played a central role in this work by publishing in 1872 his *Key to North American Birds.* This imposing reference volume enabled museum curators to unambiguously identify any bird skin lying dead on their lab table. Coues was also instrumental in publishing the AOU journal, *The Auk,* and creating its 1886 *Code of Nomenclature and Check-list of North American Birds,* which finally toppled Bonaparte's "Tower of Babel."[319]

The AOU's 1886 *Check-list* provided an officially sanctioned, carefully controlled vocabulary of Latin genus and species names for all birds known at that time in North America. Each bird also received a concise summary of its range, citations to previous ornithological books, and a single recommended English name. Unfortunately, the list didn't immediately stabilize avian nomenclature. Rather, it opened the door for incessant arguing as ornithologists tried to split apart or lump together various species and subspecies. Control over this professional discourse belonged to a junta of experts living on an axis that ran from Cambridge through New York and Philadelphia to Washington and excluded almost everyone living anywhere else.

Old Series, Vol. IX | Continuation of the Bulletin of the Nuttall Ornithological Club | New Series, Vol. I

The Auk

A Quarterly Journal of Ornithology

EDITOR.
J. A. ALLEN
ASSOCIATE EDITORS,
ELLIOTT COUES, ROBERT RIDGWAY, WILLIAM BREWSTER, AND MONTAGUE CHAMBERLAIN

VOLUME I
PUBLISHED FOR
The American Ornithologists' Union

BOSTON, Mass
ESTES & LAURIAT
1884

Volume I No. 1

Cover of the inaugural issue of the journal, *The Auk*, published by the American Ornithologists' Union since 1884. AMERICAN ORNITHOLOGISTS' UNION

However, as the country's leading publication on birds, *The Auk* was read not just by specialists but also by a growing community of amateurs around the nation who loved the birds of their own vicinity and wanted to learn more about them. They had no patience for scholastic arguments about whether to class a dead specimen under one heading or another, and found the journal's Latin nomenclature "hard to spell, hard to pronounce, hard to remember, and harder still to understand." One exasperated subscriber complained to the journal's editor, "Why don't you have it all in Latin, it would mean just as much to us."[320]

Even the AOU's English names met resistance from the general public. They were intended to reduce confusion, since vernacular names were often used inconsistently. The traditional English term "partridge," for example, was used simultaneously for the ruffed grouse, bobwhite quail, and heath hen in different parts of the country. Combined with various adjectives, it could designate six different species of game birds. Similar confusion worked in reverse, since the same bird might be called by different names in different places. The bufflehead, for example, a species of duck found throughout the entire continent, was known by at least twenty-four different local names.[321]

Displacing these traditional names was difficult. Scottish poet Edwin Muir once wrote that "with names the world was called out of the empty air, with names was built and walled." Ordinary Americans already knew what to call the birds around them, and abandoning their vocabulary for the AOU's official nomenclature meant dismantling a fundamental feature of their world. Instead of preserving the language of their elders, they were asked to conform to the practices of the scientific community.[322]

Many ordinary, uneducated people also mistrusted the methodology preached by Coues and his Victorian colleagues. Gathering data, arranging it systematically, analyzing it rationally, hypothesizing theories, and testing them scientifically may have been fine for professional ornithologists in museums and universities, but most common folk had no time or inclination for it. Just as scientific names did not instantly replace vernacular ones, scientific knowledge didn't immediately supplant faith, magic, and folklore in the minds of most nineteenth-century Americans. The next two chapters set science aside to examine how the inherited naming practices and folk beliefs of the great majority of ordinary Midwesterners prompted them to act toward birds.

8

Ahonques, Timber-Doodles, and Shitquicks

Twenty years ago I was standing in line outside Chicago's Field Museum, waiting for the doors to open and watching a flock of Canada geese nibbling in the flooded grass. A few feet behind me, a mother tried to keep her impatient children entertained. "Look at the ducks," she cajoled, pointing with an outstretched arm. "Look at those big ducks!"

As every new birder discovers, mastering the names for birds is the key to understanding everything else. You can't navigate a field guide, find information on the Web, or connect with other birders unless you speak the language. Until you understand the jargon, you're like that exasperated mom—though you can see the birds, you literally don't know what you're talking about.

Every human activity rests on this kind of common vocabulary. The ways that individual terms are constructed and how they fit together are the building blocks of every discipline. You can't understand football if you don't know the difference between a quarterback and a cornerback, and you can't understand birds if you can't tell a goose from a duck. In every field, you have to master the names in order to join the conversation.

But mastering bird names isn't easy. As a new birder, I was stumped by my first field guide. The book seemed to be arbitrarily arranged. It didn't show birds alphabetically, or by color, size, habitat, geography, or any other feature I could decipher. Instead, the birds were grouped under headings with incomprehensible Latin labels. Inside those sections, every species had not one but two more Latin names, usually based on an obscure detail that could be seen only on a dead museum specimen. The English names weren't much better. Most of them seemed to be holdovers from medieval Europe, often combined with the name of some long-dead white guy, like Bewick's wren. And then there were all the weird names I couldn't figure out at all. What the heck was a phalarope?

I remember feeling intimidated as I hunted in vain through my spotless new field guide. Something was wrong with me for not understanding how these bird names worked. After all, they had 250 years of science behind them. Who was I to question all that accumulated wisdom? I respected the names in the book as if they'd come down from heaven carved in stone, like the Ten Commandments. Standing beside a marsh with my copy of Peterson, I struggled to look up each new bird in its maze of officially sanctioned categories, which I assumed corresponded to the natural order of things.

Only later did I realize that both the names and their arrangement were artificial. Species and genera don't exist in nature; they exist in our minds. Our meticulous taxonomy doesn't reflect the world, but reflects only how modern Westerners talk about it. As Jefferson Airplane put it fifty years ago, "The human name doesn't mean shit to a tree." And the official taxonomic names don't mean much more than that to most humans, either. The terms in my field guide are not how other peoples around the globe describe nature, or how Westerners understood it before the twentieth century.

The hunters and farmers who stood beside my favorite wetland 150 years ago saw the same species that I do, but they called them by entirely different names, valued different things about them, and grouped them into different categories than those on the AOU checklist. Before that, American Indians called them by yet other names, valued them for other reasons, and deferred to other authorities.

A nineteenth-century Santee Sioux named Ohíye S'a recalled that as soon as he could walk, his grandmother "began calling my attention to natural objects. Whenever I heard the song of a bird, she would tell me what bird it came from." He was soon spending several hours a day learning about nature at the feet of his uncle, who was called Mysterious Medicine:

> When I left the tepee in the morning, he would say, "Hakadah, look closely to everything you see"; and at evening, on my return, he used

often to catechize me for an hour. . . . It was his custom to let me name all the new birds that I had seen during the day. I would name them according to the color or the shape of the bill or their song or the appearance and locality of the nest—in fact, anything about the bird that impressed me as characteristic. I made many ridiculous errors, I must admit. He then usually informed me of the correct name. Occasionally I made a hit and this he would warmly commend.[323]

In later years, Ohíye S'a praised his uncle: "Few men knew nature more thoroughly than he. . . . Nothing irritated him more than to hear some natural fact misrepresented. I have often thought that with education he might have made a Darwin or an Agassiz." His uncle insisted that Ohíye S'a get the bird names correct because tribal customs were "divinely instituted, and those in connection with the training of children were scrupulously adhered to and transmitted from one generation to another."[324]

In the last fifty years, ethnobiologists have discovered that all cultures categorize living things in essentially the same way—in a hierarchy of four to six levels ranging from the broadest through increasingly precise groups to very specific ones (e.g., animal > bird > black-crowned night heron).[325]

There are also cross-cultural similarities in how individual bird species are named. Around the globe, indigenous peoples name birds more frequently by mimicking their sounds than any other way. These onomatopoeic names usually comprise 30 to 50 percent of the bird nomenclature in societies that have had little contact with the West. Almost as many are based on a bird's plumage, shape, or unique behavior, as Ohíye S'a described.[326]

Certain phonetic principles seem to cross cultures and languages, too. Small creatures, and those that dart quickly in zigzag patterns or that display sharp angles, like warblers or rails, very often have names composed of hard consonants like *p*, *t*, or *k* surrounding high-pitched vowels like *i* or *e*. Examples in English might be "pipit" or "vireo." But large round-shaped animals and those that move slowly very often have names made from softer consonants like *sh* or *j* and lower-pitched vowels like *oo* or *ah*. An example of such an English bird name might be "loon" or "swan."[327]

As Ohíye S'a suggested, indigenous bird names are remarkably stable from one generation to the next. In 1949, Blackfoot informants easily recognized all the bird names recorded by fur trader Alexander Henry on the northern Plains around 1810, and in 1960 more than 90 percent of Inuit bird names were identical to those noted by white visitors a hundred years earlier.[328]

Names originally used in the heart of the continent are harder to discover. When Jacques Cartier sailed up the St. Lawrence River a generation after Columbus, he encountered Indians who spoke a now-extinct Iroquoian dialect. He took several of them to France during the 1530s, where a transcriber (perhaps the novelist Francois Rabelais) collected 228 words from them. Unfortunately the only bird names they recorded were goose (*sadequenda*, "big stomach") and rooster (*sahonigagoa*, "loud voice"), which is especially problematic since tribes in the St. Lawrence did not keep chickens at the time.[329]

Over the next century, the Huron Indians moved west onto the shores of the lake that bears their name today. Missionary Gabriel Sagard followed in 1623, and while living among them he began to compile the first published dictionary of any Indian language. He included about a dozen bird names but, apart from the onomatopoeic *ahonque* for the Canada goose, their derivations are obscure.

Sagard considered the Huron names "a matter of very little consequence in itself." He was "content to admire, and to praise God that in every land there is something peculiar and not to be found elsewhere." Later missionaries followed suit. Claude Allouez noted in 1677 that the Illinois Indians identified "some 40 kinds of game and birds," but he made no attempt to list or translate the names.[330]

Lay travelers also tended to overlook Indian bird names. Baron Louis-Armand Lahontan traveled extensively in the Great Lakes and Upper Mississippi Valley during the late seventeenth century and compiled a *Petit Dictionaire de la Langue des Sauvages*. Among its 350 terms and expressions in "Algonquin" (presumably a version of Ojibwe) are just two bird names: duck (*chichip*) and ruffed grouse (*pilesioue*).[331]

Luckily, the first English explorers collected a larger sample from Algonquian-speaking nations along the Atlantic coast. Their word lists appear to confirm what modern ethnozoologists find elsewhere around the globe: American Indians named birds primarily by their sounds, colors, and conspicuous behaviors.[332]

For example, most of the Powhatan bird names translated in 1585 by John Hariot in Virginia were based on observing actions, such as "digs-by-pecking" (flicker) or "bird-that-gets-himself-a-big-meal-of-seeds" (grackle). Only a third were onomatopoeic or otherwise derived from sounds, such as *chuwquareo*, which most birders will probably recognize as the red-winged blackbird.[333]

Farther north, the proportions of sight and sound were reversed. Roughly 60 percent of Narragansett terms collected by Roger Williams in New England in about 1640 referred to sounds. They consisted either of phonetic names, such as *quequecum* ("quack-quack-em") for ducks, or they described bird vocalization, such as "hoarse-criers" for herons. About 20 percent of Narragansett Indian bird names were based on visual observation. These included *sachim* for the kingbird, which, Williams explains, "The Indians give that name because of its Sachim or Princelike courage and Command over greater Birds."[334]

The only Native North American language for which zoological nomenclature has been studied in depth is that of the Iroquoian-speaking Eastern Cherokee. They still live on ancestral lands in North Carolina, in forests 2,500 feet high in the Appalachian mountains.[335]

The largest portion of Cherokee bird names are derived from hearing rather than seeing birds, as if their ancestors on the wooded mountainsides were well attuned to the sounds coming from birds hidden in trees and shrubs. Fully 42.3 percent of their names for birds phoneticize birds' songs into recognizable Cherokee syllables, in the same way that English speakers created bob-o-link. Others are purely onomatopoeic and directly echo a bird's call.

Visually-based Cherokee bird names generally come from a bird's shape rather than its colors. Thus, the tufted titmouse is called "crested-one," and the barn swallow "notched-tail." Another 20 percent of their names describe a bird's actions, such as "lucky-at-fishing" (osprey) or "climbing-up-and-down-a-round-thing" (nuthatches and creepers). Others come from conspicuous behaviors such as hunting, swimming, or flying.

Other American Indian nations appear to have labeled birds in the same ways that the Powhatan, Narragansett, and Cherokee did. Analyzing about six hundred avian terms translated in early glossaries and later anthropological reports reveals that the vast majority were derived directly from hearing or seeing birds in the field. They originated in the observer's immediate sensory experience.

Modern English bird names, in sharp contrast, obey different principles. Only 10 percent of broad avian categories refer to seeing or hearing birds (grosbeak and phoebe). Instead, two-thirds come from inherited ancient or medieval European terms (robin), to which the names of people, places, or other abstract terms were often added (Lucy's warbler). About half of the specific names in modern field guides refer directly to visual features (indigo bunting) and just 2 percent are based on a bird's call or song (trumpeter swan). The bird names used most frequently in the Midwest today generally derive from abstract concepts rather than from direct experience of sights and sounds.[336]

When I try to identify a tiny flycatcher high above me in the leaves, I'm imposing a conceptual structure on a disorderly external world. When the bird darts off and I mutter, "Damn! Which one was it?," I unconsciously assume that the external world is chaotic and that by assigning the correct scientific name to part of it, I will be able to impose order. Jesuit missionaries living here four hundred years ago did the same thing. They saw themselves engaged in a "perennial struggle between the ordering mind and the recalcitrant phenomenon." The unconscious assumption reinforced by language is that humans stand apart, are separate, from nature.[337]

But indigenous peoples find this perspective silly, because to them nature is not chaotic and imperfect; humans are. Anthropologist Hamilton Tyler describes how American Indian names for birds, animals, and plants are "not primarily designed to make sense of an outer world, because the natural world is assumed to *be* the sense." In traditional indigenous societies, calling plants, animals, and birds by ancient names based on direct sensory experiences helps to keep humans closely connected to nature. For modern Americans, naming natural things with inherited European terms sets them apart from nature.[338]

Although most American Indian bird names were derived from subjective sensory experiences like hearing a bird's song or watching its behavior, they still functioned as symbols, just as Western terms do. In a similar way, Chinese characters are derived from images, but over three thousand years of use, their visual content has been largely eclipsed. Seeing a character on a page, Chinese speakers don't see a picture, they recognize a concept. A similar transformation presumably happened to "digs-by-pecking" and *quequecum*. In everyday speech, the names are a sequence of phonemes that represent an idea. Yet at the same time, they embody an immediate personal experience of sights and sounds.

American Indian bird names also contain meanings that link to other aspects of their cultures. When Ohíye S'a and Mysterious Medicine used a bird's name, they embedded themselves inside something much more powerful than a scientific convention—they embedded themselves inside their community's religion and social structure. For thousands of years, birds have featured prominently in American Indian creation narratives, medical practices, religious ceremonies, clan identities, food habits, dress, ornamentation, and art. Allusions to birds permeate the entire cultural experience of American Indians. Tyler described how, when using the term *ta'wa ma'na* ("sun maiden") for the yellow-headed blackbird, a Hopi speaker also conjured up a symbol of the north, an offering to the sun (one of the agents that brings summer each year), and a tribal clan.[339]

This intimate connection between culture and nature is found among indigenous societies around the globe. Ethno-ornithologists Sonia Tidemann and Andrew Gosler describe it as "a seamless continuity of perception between the physical, spiritual and cultural environments of people. In this perception, humans do not stand distinct from their environments or other animals but are an integral part of the former interacting on equal terms with the latter." Henry Thoreau described it more poetically in 1841: "The Indian stands free and unconstrained in Nature, is her inhabitant and not her guest, and wears her easily and gracefully."[340]

So Mysterious Medicine and Ohíye S'a placed as much confidence in the traditional Santee bird names as modern Westerners place in those printed in today's field guides. And perhaps with better reason. After all, the Santee names came from the creator of their world, while the modern ones just come from the American Ornithologists' Union.

When the French first sailed up the St. Lawrence in the sixteenth century, they could use only existing European names for the birds they encountered. For example, at the end of June 1534, Jacques Cartier reported seeing "canaries, linnets, nightingales, sparrows and other birds, the same as in France." One could argue that perhaps the first white observers did not see American birds at all. Instead, they saw what they expected to see, European birds, and described them in European terms.[341]

But as they headed farther west into the heart of the continent, the French gradually realized that their inherited bird names didn't always fit American birds. Louis Hennepin, who accompanied La Salle in the Upper Mississippi Valley about 1680, reported that in Michigan, "Our Men brought several other Beasts and Birds, whose Names are unknown to us, but they are extraordinarily relishing."[342]

One strategy for discussing these unknown species was to stretch a European name to cover the new circumstances. For example, in France the name for the common wild goose was *outarde*, a term originating with Pliny that eventually became Anglicized as bustard. In America, it was used to refer to the Canada goose, an entirely different species from the Old World *outarde*. In the same way, the French word *ortolan*, which referred to a specific European game-bird highly regarded for its delicate taste, became to French Canadians "a common epithet . . . for all the lesser feathered race that are eatable, and whose real names they are unacquainted with."[343]

Some American birds, however, were so unusual that traditional French names couldn't be stretched to cover them, and observers had no choice but to invent new ones. About half the bird names in the earliest French accounts of North America occur in the sole French ornithology of the day, Pierre Belon's 1555 *L'Histoire de la Nature des Oyseaux*. The others were coined on the spot or borrowed from local American Indian languages. This was the case with a small bird that had no European equivalent and whose striking habits and iridescent colors fascinated early observers.

"We saw too a tiny bird," begins the first French account of the ruby-throated hummingbird, from 1606, "no bigger, including all its feathers, than an almond in its shell. It flits about like a butterfly and lives only on flowers, just like the honey bees. . . . The natives of the land [Micmac Indians] call it in their language *nirido*." And so, therefore, did the French, adopting the American Indian name because they had no pre-existing term. It appears as *nirido*, *nerido*, and *niridau* in early French texts; the modern phonetic rendering from Micmac is *militaw*.[344]

Other observers coined more descriptive names for the tiny bird. At various times it was the "Bird of Heaven," "flower-bird" and "bird-fly" (*l'oiseau-mouche*). The last was used by two Jesuit missionaries, Paul Le Jeune in the St. Lawrence Valley and Gabriel Sagard in the country of the Hurons. It ultimately became the standard French Canadian term until modern times.

The same thing happened as survivors of La Salle's ill-fated 1685 expedition headed north up the Mississippi from the Gulf of Mexico. When they encountered the first pelican, they called it "le grand gosier, or the great Gullet, because it has a very large one; another . . . we called the Spatula [spoonbill], because its Beak is shap'd like one."[345]

Two generations later, in 1734, French settlers in the lower Mississippi gave a conspicuous red bird the name we still use today. "The Cardinal," wrote a visitor, "owes its name to the bright red of the feathers, and to a little cowl on the hind part of the head, which resembles that of the bishop's ornament, called a camail." With one bird already named after ecclesiastical garb, they called a second species "Bishop" (the indigo bunting) and elevated a third (the painted bunting) to the head of the triumvirate as "Pope," a name that persisted well into the next century. According to an eighteenth-century French settler in the Mississippi Valley, "It got that name perhaps because its colour makes it look somewhat old, and none but old men are promoted to that dignity; or because its notes are soft, feeble and rare; or lastly, because they wanted a bird of that name in the colony, having two other kinds named cardinals and bishops."[346]

Other French names for American birds were purely descriptive, such as *canard branchu* for the wood duck. "It perches in the trees," noted traveler C. C. Robin at the turn of the nineteenth century, "so they call it here, with good reason, the wood duck." Other examples include *oiseau puant* ("stinky bird") for the turkey vulture, from its habit of eating decaying carcasses, and *aigle a tête blanche* for the bald eagle, from its white head.[347]

This habit of local residents inventing their own bird names naturally led to great confusion, especially as multitudes of new immigrants fanned out across the Midwest in the opening years of the nineteenth century. Audubon, strolling down to the riverfront from his Louisville shop, would have seen a steady stream of people floating down the Ohio into the Mississippi Valley. By 1815, this initial trickle had grown into a flood of farmers, Indian traders, merchants, mechanics, hunters, miners, and enslaved people heading west. Many settled near the river in southern Ohio, Indiana, or Illinois, but others continued across the Mississippi onto the prairies of Missouri or north to the lead mines of Iowa and Wisconsin. Thousands of others left New England and New York, floated along the Erie Canal, launched into the Great Lakes, and came to rest in the fertile valleys that fed them. Between 1800 and 1825, more than a million new settlers migrated into the "Old Northwest."

Few, if any, of these pioneers could write a book. Most, perhaps, couldn't even read one. They'd never encountered debates about avian classification in the journal of the Philadelphia Academy of Natural Sciences and never heard of Latin binomials. Instead, they talked about birds in ways that had been passed orally from generation to generation. In their fields and around their hearths, rural and working-class Americans discussed nature on their own terms.

Many of their bird names were passed down over the centuries from the first Europeans to land on this side of the Atlantic, who attached Old World names to New World birds. The first thirty English accounts to mention American birds employed terms from Elizabethan ornithological works 90 percent of the time, even when they didn't quite fit.

For example, careful English observers recognized from the start that the ubiquitous red-breasted North American thrush was not identical to the "robyn redbreast" back home, and tried to make the distinction clear by giving it new names such as "Fieldfare of Carolina" or "red-breasted thrush." Yet as early as 1748 English residents were doing what the French had done with *outarde* and *ortolan*: everyone in the mid-Atlantic region called it "robin-red-breast," even though they knew full well that "it is a very different bird from that which in England bears the same name."

By the time of the American Revolution, the old English name was so securely fixed to the New World bird that Sir Charles Blagden could write from Rhode Island, "The *Turdus migratorius* . . . is as familiar to every child in America under the name Robin, as the little bird its namesake is in England." And when Alexander Wilson wrote the first comprehensive work on American birds twenty-five years later, he had no choice but to use the ubiquitous common name.[348]

Unlike French and English settlers, however, the Pennsylvania Germans didn't rely heavily on inherited European names for the birds that surrounded them. They initially settled in counties just west of Philadelphia, but starting early in the nineteenth century gradually spread out across Ohio, Indiana, and Illinois and into the Mississippi Valley. Everywhere they went, these Amish and Mennonite Germans erected small homes with large barns, fenced their pastures, manured their fields, rotated their crops, and conserved their woodlots. When their heritage was threatened by the encroachment of mainstream, English-speaking society one hundred years ago, the Pennsylvania Germans deliberately collected and printed their traditional language and folkways in order to preserve them.

"Birds had a place in that life," Pennsylvania historian William Rupp later wrote, "of greater import and wider significance than most of us have so far been able to imagine." Although at first they spoke the dialects of their various homelands in Switzerland or Germany, they gradually evolved a unique language that had German roots but sprouted American shoots. Their bird names are as different from German ones as are the birds themselves. Instead of forcing Old World names onto New World birds, they coined original names during their daily life in barns, fields, and pastures. And their names for birds express an intimacy with nature similar to that conveyed by American Indian names.[349]

Traditional German names account for just 14 percent of the terms that Pennsylvania Germans applied to birds, and these were usually broad terms such as *kranich* (cranes) or *eil* (owl), rather than specific ones. Another 6 percent were cognates from English, such as *bussard* for buzzard (turkey vulture). The remaining 80 percent of bird names used by Pennsylvania German farmers were coined on the spot from sights and sounds, like American Indian ones.

Two-thirds were based on visual features. Nearly half of these were derived either from the bird's color, such as *rotschwenischer hinkelwoi* (literally "red-tail-feathered chicken-hawk" for the red-tailed hawk) or from a bird's eating or hunting habits (*kaerschevoggel* or "cherry bird" for the cedar waxwing). Another third were based on behavior (such as *faulenser* or "lazybones" for the nighthawk, from its lying around all day) or on habitat (*schwammstaare* or "swamp blackbird" for the redwing). Other sight-based names came from observing birds' migration habits (*schneegans* or "snow goose" from the Canada goose's habit of arriving at the start of winter), from their size (*mausevoggel* or "mouse bird" for the house wren), or from their flight behavior (*schdoosswoi* or "darting hawk" for the sharp-shin).

About 12 percent of Pennsylvania German bird names derived from hearing a bird's call, suggesting that farmers relied more on their eyes than their ears while outdoors. The majority of sound-based names were simply onomatopoeic (like *schewink* for the towhee), but some were descriptive, such as *lufthutschel* or "flying-whinnier" for the common loon, and *katzevoggel* for the catbird. A handful of names were even based on olfactory experience, including *schtinkend* ("stinky duck") for the coot.

Another nineteenth-century community whose daily lives were immersed in nature and whose unique bird names have been preserved were professional hunters who killed and sold game to urban markets. By the middle of the nineteenth century, railroads had enabled rural gunners to kill and ship birds to customers in cities. Professional hunters at prairie ponds or along the Great Lakes could make a living harvesting birds for markets in Cleveland or Detroit, and farmers in Wisconsin and Iowa could take days off to kill birds that would later be eaten as far away as New York.

The bird names used by these hunters varied from region to region. Foreign names persisted in French-speaking parts of Canada and Louisiana, and among German settlers in Wisconsin and Missouri. These differed from the English names that spread west with Yankee settlers or came north with former slaves. In 1888, Connecticut sportsman Gurdon Trumbull collected several hundred vernacular bird names used by rural residents in the eastern half of the country in his marvelously titled book, *Names and Portraits of Birds Which Interest Gunners, with Descriptions in Language Understanded of the People*.

In his introduction, Trumbull explained:

> Many of those English names which perhaps we all ought to adopt, such as Hooded Merganser, Hudsonian Godwit, Bartramian Sandpiper, Pectoral Sandpiper, etc., are used about as little by the inhabitants of the United States generally as the strictly scientific names. . . . our gunners have, as a rule, proved themselves a very conservative class, continuing the bird names of their forefathers persistently, despite the teachings and sneers of scientists and book-learned sports-

> men. Many of these names, probably, appear now for the first time in print, yet few are of recent origin; and though some may be a little time-worn, they are time-honored, and as familiar in certain localities as "cow," "dog," and "cat." . . . Names which appear to us absurdly grotesque and outlandish are mediums of communication between men as wise as ourselves, though educated in a different school, and the homely nomenclature of those who shoot, not alone for sport, but for their daily bread, should command respect.[350]

Thirty years later, Trumbull's obscure book took on a new life when the United States, Canada, and Great Britain signed the Migratory Bird Treaty to protect American birds. This 1918 treaty identified birds by their scientific Latin names and official English ones, but those meant nothing at all to semiliterate gunners in Missouri swamps or Indiana uplands. A translation was needed from the AOU checklist to local parlance.

So a US Biological Survey ornithologist named Waldo Lee McAtee compiled a list of bird names actually used in the field. His goal was to educate the hunters and to help game wardens enforce the treaty. McAtee sent a questionnaire to naturalists and hunters around the country, adding to their responses all the terms in Trumbull's book or that he had personally heard in the field. The result was a ninety-five-page monograph with more than 1,800 colloquial bird names—an oral tradition created over several generations that was almost entirely independent of the scholarly nomenclature used by academic ornithologists.[351]

About 83 percent of the bird names reported by McAtee's market gunners were based on sight or sound, like virtually all American Indian names and 80 percent of Pennsylvania German farmers' (but just 10 percent of the first European explorers' bird names). Most of these were based either on plumage (such as "black-breasted sandpiper," used in Wisconsin for the dunlin) or upon the habitat where the birds were seen ("marsh bluebill," used in Michigan for ring-necked duck). Others came from observations of behaviors, such as "teeter-bob" in Indiana for the spotted sandpiper.[352]

Gunners rarely based their bird names on hearing (just 5 percent) and nearly all of those described a bird's sound ("whistling plover" used in Ohio for the upland plover) rather than directly mimicking its call; an exception was the Ohio name "weet-weet" for the solitary sandpiper. A few even referred to how a bird tasted, like "butterball," used for several species of ducks.[353]

Just 17 percent of local hunters' names for birds were abstract rather than sensory, and they were mostly survivals from archaic English or French terms, such as *brant*. A few were quite fanciful, such as "hell-diver" for the hooded merganser in Wisconsin or "timber-doodle" for the woodcock (used in the Great Lakes).[354]

About 2 percent of the traditional vernacular terms were explicitly pornographic or scatological, but prevailing moral codes prevented Trumbull and McAtee from publishing them in commercial or government publications. Luckily for us, McAtee resorted to privately printed pamphlets in order to keep them from vanishing forever.

McAtee was once described by a colleague as "the most honest man I have ever known," and he had no patience with genteel circumlocutions. He considered blunt language "a sign of the robust heartiness that is an important part of the make-up of every well-endowed human being" and viewed his linguistic work as a "battle that must always be waged against obscurantists, who would restrict the domain of language, and hypocrites who pretend that certain classes of words do not exist."[355]

Bird names that could not be printed one hundred years ago were often derived from a bird's excretion, the various herons in particular being very widely known by excretory names. As early as the 1770s, New Englanders called the great blue heron "shite-poke," the first half of the word being an early equivalent of *shit* and the second meaning *bag* (as in the old proverb about not buying a "pig in a poke"). The intended meaning was derived from the great blue heron's conspicuous habit of evacuating its bowels when taking off in fright.

McAtee found thirty-four variant spellings or polite euphemisms for "shit-bag" used all across North America, as well as Spanish, German, and French equivalents. The Pennsylvania German version was *sheidpoke*. "Shidpoke," "shitepoke," and the sanitized version "slough poke" were other Midwestern variants. In Florida, when residents saw the heron emblem on the uniforms of National Audubon Society wardens they quickly dubbed the organization "the Shitquick Society."[356]

Herons were not the only such case. McAtee wrote that "there is no excuse save politeness for calling the sound accompanying the dive of the nighthawk a 'boom,'" since it was "perfectly imitated by a voluminous and lusty dry fart playing across a throbbing margin of anal membrane." For this reason Pennsylvania German farmers called the nighthawk *fatzvogel* or *luft-fatzel* ("fart bird" and "air farter"), and Finnish residents in Minnesota employed *pieru lintu* ("fart bird").[357]

Another class of indecent bird names derived from the action of a bird wiggling its hindquarters. Spotted and solitary sandpipers that, in the polite words of my field guide, "nod and teeter constantly," were known as "teeter-butt" in Iowa and "teeter-ass" throughout the Great Lakes. The latter was "the only common name these two species are known by" in much of the region, despite appearing in print rarely, if at all.[358]

Working-class and immigrant observers freely applied bird names like these because they didn't share the "polite" manners of the educated, English-speaking elite who controlled the bird names that appeared in books. Sitting in the parlors of Philadelphia in 1810, Alexander Wilson and George Ord would not have called the great blue heron a "shitpoke" because the conventions of their class shunned such crass language. But outside their window, people who survived by physical labor and lived in close quarters with little privacy were on different terms with the coarser facts of life.

For most of his career, McAtee kept a running account of the avian terms he encountered in the field or while reading widely in the history of ornithology. His decades-long project yielded ninety file drawers of paper slips containing sixty-seven thousand common names of birds. He turned these into a 1,700-page dictionary arranged in the same order as the AOU's *Check-list*. Each entry consisted of a bird's scientific name followed by all its vernacular ones, and a second volume was to have included indexes and cross-references. The University of Chicago toyed with publishing it but eventually decided against bringing out two ponderous, expensive tomes that would appeal only to a tiny market of specialists.[359]

By the time McAtee died in 1962, conservation laws had rendered market gunners as extinct as passenger pigeons. America's rural population had diminished, and the officially sanctioned bird names had been widely disseminated. The event that finally lodged scientific bird names in the minds of the general public was the publication of Roger Tory Peterson's 1934 *Field Guide to the Birds*.

The book's two great assets were its size (it fit easily into a coat pocket) and its diagnostic hints. Earlier reference books, such as Nuttall's 1832 *Manual* and Coues's 1872 *Key*, had assumed that readers were examining a dead specimen on a laboratory table. Peterson's text and illustrations were the first to emphasize specific markings that could be clearly seen from a distance on a live bird and that distinguished each species from all others. In addition, every bird was identified by its AOU-sanctioned Latin binomial and official English name, and the whole book was arranged in AOU taxonomical order.

At least four publishers turned down Peterson's manuscript before Houghton Mifflin, also skeptical, agreed to print two thousand copies. These sold out in two weeks, and after 1934 no recreational bird-watcher or academic ornithologist wanted to go into the field without one. Peterson's guide was reprinted in 1939, a companion volume covering western birds was issued in 1941, and since then the pair have remained continually in print, selling more than seven million copies and spawning an entire genre, which recently morphed into mobile apps.[360]

Beginning in 1934, virtually everyone who cared about birds, laypeople and professionals alike, mastered the AOU bird names because they used Peterson and its successors in the field. Today, bird-watchers standing by a wetland will call a ruddy duck by whatever name their field guide or smartphone app tells them to call it. They will never know that to their great-grandparents it may have been a "booby coot," "dip-tail diver," "sleepy-head," "water-partridge," or "paddy-whack," not to mention its 150 other local names.[361]

9

Feathers, Fetishes, and Fables

"Come on, Mike! Come pull the wishbone!"

It must have been the fall of 1957. I can still see my grandmother in the back kitchen of her tiny New Hampshire house calling me to join my big brother in pulling apart the Thanksgiving turkey's wishbone. I must have been quite young, since she had to explain to me that the bone would break into a big piece and a smaller one, and that the person who got the larger piece would have his wish come true.

I was puzzled, even at age four, at how this could possibly be true. My dad was an engineer who assumed science would save the world and considered religion a mental illness. He'd already instilled a no-nonsense approach to life in me and my brother. After all, this was the age of atom bombs, interstate highways, and *Sputnik*. Wishing on a turkey bone seemed like nonsense to me. But I did what I was told.

I doubt my grandmother had ever believed that wishes come true. She was a stoic, working-class Yankee with strong opinions and few illusions. But it's possible that her own grandparents, uneducated nineteenth-century farmers in the Connecticut Valley, had believed it. Science didn't penetrate very far into their lives; I once heard her father, my great-grandfather, worry that NASA's first rockets would knock down the moon. Family stories, nursery rhymes, and maxims passed down verbally for generations probably carried more weight with my ancestors than book learning.

By grasping the turkey bone in my tiny fist, I was perpetuating a vanished world that my grandmother could clearly remember, much as I can recall that Thanksgiving Day sixty years ago. Reason and science may be common sense today, but when her grandparents were young, they were the private domain of educated specialists. The vast majority of people got by perfectly well without them.

Between 1790 and 1840, millions of Yankees like my great-great-grandparents—as well as French Canadians, Pennsylvania Germans, and immigrants just off the boat from Europe—crossed over the Appalachian Mountains or sailed up the Great Lakes into the heart of the continent. Uncounted African American slaves migrated north across the Ohio River or up the Mississippi with (or liberated from) their masters. They all rubbed shoulders with the Native peoples who had been living here for thousands of years.

Along with their rifles, axes, and plows, these newcomers carried knowledge acquired in the woods and fields rather than in books and classrooms. They preserved it not in libraries but in their memories, and they shared it not in print but in proverbs, customs, and stories passed aloud from generation to generation.

Rural folk trusted, for example, that if a cardinal crossed their path from left to right, good luck would follow; but if it crossed the other way, bad news was coming. Many thought that swallowing a raw dove's heart could make someone fall in love with them. They believed that witches sometimes occupied the bodies of turkey vultures and had to be either placated or avoided. An owl hooting on their rooftop signaled approaching death or misfortune—unless they took off some clothes and put them back on inside out. Superstitions like these filled the minds of exponentially more Americans than did the conclusions of professional ornithologists.[362]

Most of this traditional folk knowledge about birds went extinct with the passenger pigeon about a century ago, driven out of people's heads by popular education and mass media. By the 1860s, public schools began teaching scientific explanations to millions of young minds, loosening the grip of their grandparents' oral traditions. New printing technology spawned newspapers in every small town, and telegraph lines carried Associated Press stories into their pages. By the end of the nineteenth century, Midwesterners whose parents had lived in relative isolation on the frontier could read about the latest scientific discoveries in their kitchens and parlors. Their parents' folk knowledge lost its authority, became merely quaint or was held up for ridicule, and was at risk of being forgotten altogether.

So at the turn of the twentieth century, a handful of collectors deliberately tried to gather these vanishing notions. A small corps of folklorists, linguists, and antiquarians captured the popular sayings, superstitions, customs, and stories that had been shared for generations among working-class and rural people, including many ideas about birds. They distributed questionnaires, interviewed elderly pioneers, transcribed conversations, recorded recollections, and took down stories.[363]

It's impossible to know how seriously people took this folklore. Some sayings may have been recited as playful proverbs by children, like nursery rhymes. Some may have been the hollow shells of once-powerful rituals, akin to my pulling the wishbone in 1957. But we know that others were taken so seriously that people died by following their prescriptions. Although most of the examples given below are associated with a specific ethnic community, this doesn't mean that every Pennsylvania German or black Missourian subscribed to them. In every community, some people probably thought such beliefs were nonsense at the same time that others swore by them.

Taken as a whole, though, folklore about birds reveals ways of thinking and acting that were embraced by the majority of Americans before the twentieth century, when reason and science became mainstream. And folk beliefs were effective at calming their fears. The scientific causes of natural disasters, economic collapses, illnesses, injuries, and personal misfortunes were beyond most people's comprehension a century ago. Time-honored maxims, traditional customs, and familiar tales helped our ancestors make sense of a world that was unpredictable at best, and often hostile and frightening.

Ornithology is two centuries old, but American folklore about birds originated many centuries ago, reaching deep into the roots of our culture in Europe, Africa, and archaeological time. Any history of people and birds must take it into account.

People around the globe have always passed advice to new generations in the form of proverbs. When they encapsulated a common experience in a memorable way, using rhyme or alliteration or parallelism, these proverbs survived many generations as oral tradition and in print.

Folklorists recorded more than fifteen thousand proverbs used by Americans before 1880. Birds played a minor role in this proverbial wisdom—just a few dozen proverbs that refer to birds are found in the standard scholarly collections—but some became so widespread that they are still familiar to us today.[364]

"Birds of a feather flock together" is a notion that goes back to Homer and came to America in 1674 with Massachusetts colonists. "To kill two birds with one stone" was used first in America by Reverend Cotton Mather in his 1692 book on witchcraft, *Wonders of the Invisible World*. "A bird in the hand is worth two in the bush" would have been an obvious truth to every frontier settler roaming the fields with a flintlock in search of dinner. It first appeared in America in about 1700 in a poem by Mather's fellow minister, Edward Taylor.[365]

Proverbs often epitomized the key values of a community. Just as the Protestant work ethic was captured in the sayings of Mather and Taylor, the power relations of Southern slave life were expressed in proverbs recorded by Joel Chandler Harris just after the Civil War: "Pullet can't roost too high for de owl"; "Settin' hens don't hanker arter fresh aigs"; "Hit take two birds fer to make a nes'"; and "Rooster makes mo' racket dan de hin w'at lay de aig."[366]

Charming as they may seem to contemporary readers, these proverbs aren't really about birds; they simply employ birds to express truths about gender and power relations. A separate, much larger body of sayings and maxims captured by folklorists a century ago preserved how ordinary people thought about the birds that surrounded them.

For instance, many sayings about birds were related to the weather. Until quite recently, historically speaking, the majority of Americans spent the bulk of each day outdoors. The 1920 US Census is the first to show more people living in urban areas than on farms.[367] Before then, most people lived in the country, growing crops, raising livestock, harvesting trees, and hunting game. Sudden weather changes could mean catastrophe for rural residents, and birds were one way that people predicted them.

Frontier settlers watched how birds reacted to changes in temperature, air pressure, and humidity, noting that "wild geese fly high in pleasant weather and low in bad weather," "gulls will soar to lofty heights and circling round utter shrill cries before a storm," and "when the swallows fly high over the fields, it is a sign of favorable weather; if they fly low over the meadows and ponds, it is a sign of rain." A nineteenth-century meteorologist claimed that "when the barometer is high the air is heavier and denser and has more sustaining capacity, and birds are therefore able to fly or soar high with less effort than would be required at times when the barometer is low and the air less dense."[368]

Every barnyard and most urban neighborhoods hosted chickens until the twentieth century, and they were widely thought to signal changes in the weather. It was believed that "if the chickens run and seek shelter when it begins to rain, the rain will not last very long," and that "when chickens light on fences during rain to plume themselves, it will soon clear." Some people thought they could also predict the next day's weather: "The rooster who crows when he goes to bed will get up with a wet head."[369]

Birds also marked the changing seasons in phrases like, "When summer birds take their flight, the summer goes with them," and "The northward flight of the Canada geese in late winter and early spring is a sign of warmer weather." Some seasonal proverbs sound like common sense but reach dubious conclusions from their premises, such as, "If the chickens grow feathers on their lower legs and feet, it is an indication of a severe winter," and "If there be ice in November that will bear a duck / there will be nothing thereafter but sleet and muck."[370]

In addition to predicting the weather, birds were also thought to cure a wide variety of human ailments. These home remedies probably evolved by trial and error through many generations and, in the absence of professional medicine, most likely provided some relief to the afflicted. But others simply sound crazy—you don't want to try any of these at home.

Various birds were thought to relieve muscle and joint pain if their blood or carcass was applied to the injured area: "A live pigeon torn in two and applied to the parts of the body affected with acute rheumatism will cure it." In Illinois, Michigan, and Missouri, snakebites were treated "by applying to the bite the raw and bleeding surface of the flesh of a fowl that has been stunned. Sometimes several are used in succession." The patient had to take care when killing a bird, though, because "If a chicken die in one's hands, it causes palsy." And killing a defenseless mourning dove might cause a boil, carbuncle, or other skin eruption.[371]

To cure rheumatism, popular folk wisdom advised German settlers to "shoot a carrion bird or turkey buzzard [*einen Aas Vogel oder Torke Bohser*], cut off the feathers with shears and hang the carcass to the fire so that it will roast; catch the fat drippings and then smear your limbs as long as you feel any pain." For rheumatism, asthma, and epilepsy, rural African Americans believed that "two wing feathers of the buzzard are effective if burned under the nose and the smoke inhaled."[372]

Birds also nurtured frontier romances. Young German men were told to "carry the heart of an owl in your pocket if you want to have the admiration of the girls." And African American suitors were advised that "a white dove's heart swallowed raw with the point downward" would win over the object of their affection.

No bird is mentioned more often in connection with love than the cardinal. In the Ohio Valley, settlers believed that "after you see a redbird, the next person that you see will be your sweetheart" and "if you throw a kiss to a redbird, he will take it to your sweetheart." African Americans believed that the process was a bit more complicated:

> Should a girl see a redbird and name it after her sweetheart, she will see him before sunset; or should she throw a kiss (or nine kisses) at the bird, she will see her sweetheart at that same time the next day. Watch the direction in which the redbird flies; your lover is surely in that direction. If the bird flies in your front yard, your lover is coming to see you that same night.

Black residents in the Ohio Valley also believed that a flock of birds flying over the church at the time of a wedding signified a happy future for the married couple. And birds' nesting was also linked with marriage: "If you find a bird's nest empty, you will never marry. If it has one egg in it, you will marry within one year; if two eggs, two years; et cetera."[373]

Similarly, many people believed in birds' ability to bring them good luck. It was widely thought that, if two people made wishes while pulling apart the breastbone of a roasted chicken or turkey, the one ending up with the longest piece would have their wish come true. According to the Germans, "The one getting the shovel or long piece will live longer than the person getting the mattock or short piece, for he has the shovel with which to help bury the other." Another bit of chicken anatomy could also create good fortune: "To swallow a raw chicken-heart brings good luck." African Americans believed that wearing the foot of a blue jay, like carrying a rabbit's foot, would do the same.[374]

African Americans also thought cardinals had special powers to make wishes come true: "Wishing on the redbird is done sometimes simply upon seeing him but generally it is necessary to throw a kiss (or nine kisses or five kisses each containing six wishes) for your wish to come true." Blackbirds, bluebirds, doves, vultures, and whip-poor-wills all had the power to make wishes come true, too, though sometimes elaborate rituals had to be performed. Some black families reported that one's wish could be guaranteed only "by turning over three times in bed on hearing the first whippoorwill of the year, or as the more meticulous say, by going outdoors and rolling over three times in the direction of the call. Some substitute the mourning dove for the whippoorwill, while the lazy man simply makes his wish on the first whippoorwill without the attendant gymnastics."[375]

Many folk beliefs about birds reflected unconscious fears or desires. In northern Ohio, any bird flying into a home was considered a harbinger of death; in many places, a bird merely trying to get in at the window signaled approaching demise. A prophecy of death in Kentucky's Blue Grass country was caused by a picture falling down at the same time a bird entered one's home, and in Kansas by a bird landing on someone's head. Crows, doves, whip-poor-wills, and woodpeckers were all associated with fatalities in various places. In African American communities, some people took a woodpecker tapping on the roof as "a sign of the screws being bored in the coffin of one of the inmates."[376]

However, no birds were so closely associated with the Grim Reaper as owls. One American folklorist speculated that "this almost universal superstition is doubtless due to the nocturnal habits of the bird and his strange half-human cry, and possibly represents a remnant of the belief of the late Middle Ages that such birds were evil spirits coming to devour the souls of the dying." An owl hooting on one's roof was the classic example, but seeing one at the window could be equally fatal. In some locations, the notion was subtly nuanced: "Some say that the omen holds only when the owl hoots in the daytime; others apply it only to the house of a sick person; while still others claim the hoot of an old owl denotes the death of an old man and that of a young owl the death of a child."

The victim could sometimes be saved by heating a poker or other metal until it became red-hot in the fireplace, or by turning one's clothes inside out. African Americans had a complete repertoire of defense:

> To avoid the ill luck or to make the owl hush up, stick a knife in wood; squeeze your wrist (this will choke the owl to death); turn your pockets wrongside out . . . ; turn some shoes upside down under the bed so that his claws cannot grasp the limb; put salt or a shovel (or a horseshoe) in the fire to burn his tongue (or strangle him); or tie a hard knot in the (right-hand) corner of your bed sheet, handkerchief or apron to choke him to death.
>
> These represent the main ways of stopping the hooting of the owl but there are almost countless variations. Lock your little fingers together and pull hard; put a hairpin over the lamp chimney (possibly a variant of the iron thrust into the fire); turn your socks or clothes inside out; put one shoe with the toe pointing under the bed and the other shoe with the toe pointing out; lay a broom (an old-fashioned sedge broom) across the door; cross two sticks and put them in back of the chimney to break the owl's neck.[377]

But birds could also be good omens. Residents of Kansas believed that "if a white crow or a white blackbird fly over one's head, very good luck will attend that person," and in Kentucky, hearing a passenger pigeon cry on the first day of the year indicated a bountiful harvest that fall. "Notice when you hear the first turtle-dove of spring," Kansans told each other; "if you are walking or standing, you will be well; if sitting, you will be sick; if lying down, you will die." Germans said, "Shake your purse when you hear the call of the whip-poor-will for the first time in spring and you will always have money in it during the ensuing year." For rural families on isolated farms, the arrivals of visitors were especially welcome events, and cardinals, robins, woodpeckers, and turkey vultures flying before the door or above the house could all signal a happy, unexpected visit from a friend.[378]

Many of these beliefs take us from the realm of credibility—bird behavior prompted by changing temperatures or air pressure—into that of the supernatural. Although every ethnic group in the United States had good-luck charms and esoteric superstitions of some sort, the occult beliefs and rites of African Americans are the best documented in the Midwest.

In 1830, just twenty-six thousand black residents lived in the entire region, nearly all of them enslaved people in Missouri. But during the next generation, large numbers of African Americans escaped from the South into Ohio, Indiana, and Illinois or were taken by slave owners as far north as Wisconsin and Minnesota. After the Civil War, formerly enslaved people settled in cities all across the region. Between 1830 and 1880, the Midwest's black population grew more than tenfold to nearly four hundred thousand, and by 1900, more than half a million African Americans lived in the Midwest.[379]

They carried north with them many beliefs that dated back generations. During the 1890s, century-old voodoo practices that originated in Louisiana were found thriving in Missouri, and in the 1920s, esoteric ceremonies from the southern Black Belt were documented in the black community of Cleveland, Ohio. As late as 1953, folklorist Richard Dorson collected more than one thousand folktales in Michigan from African Americans who had moved from the South with their families since the 1880s.[380]

Enslaved people who were brought to the United States from Africa and their immediate descendants inhabited a supernatural world similar in many ways to the shamanist cosmos of American Indians. To them, the world experienced through the senses and explained by reasoning was just the superficial surface of a universe alive with spiritual powers. Health and disease, good luck and bad, success in love or business, and most other events could be influenced by these invisible forces. "Hoodoo" or "voodoo" experts, both men and women, were believed to intervene in the supernatural realms to produce results in everyday human affairs.[381]

Anthropologists have traced these early African American beliefs to different African sources. In the French- and Spanish-speaking areas of the Gulf Coast and Lower Mississippi Valley, West African religions of the Fon, Yoruba, and Kongo peoples mixed during the late eighteenth century with the folk Catholicism of European residents and shamanism of American Indians. In English-speaking areas on the Atlantic coast, on the other hand, the Protestant religion practiced by both masters and indentured servants mingled with the ideas of predominantly Kongo slaves.

During the nineteenth and early twentieth centuries, African Americans from both regions spread across the South, west into Missouri, up the Mississippi Valley, and north into the Great Lakes states, creating a spectacularly rich heritage of magical beliefs and occult practices. "I've been all over the country," an educated black resident of New Orleans reported in the 1940s, "and I've seen signs of Voodoo almost everywhere, anywhere people of my race live. You can always find it. Of course, lots of white people don't know anything about it, but we always know. Anywhere they go, my people know the signs."[382]

Birds played important roles in underground voodoo culture. Witches, for example, were known to "change their appearance to all forms, from gnats to horses, but especially prefer the shape of large birds such as the buzzard." Good spirits, on the other hand, "appear as white doves, men and children (at times with wings), or else look like mist or clouds. The air from these good spirits will cure sickness. They are able to fly high, but the evil spirits remain close to the ground and lead men into ditches or briar-patches." Normal mortals could not see witches or evil spirits and instead had "to rely to some extent upon screech owls and dogs, which creatures always make a noise when a witch is approaching."[383]

Hoodoo adepts had the power to abandon their bodies and take animal forms. Owls were an especially popular host and were widely feared. Depending on the circumstances, to hear an owl signaled death, bad luck, or approaching danger, and to imitate or mock an owl was certain to bring punishment or cause one's house to burn down. While the hoodoo priest or priestess was in animal form, the devil might inhabit their human bodies in order to move unseen among mortals.[384]

Sickness, misfortune, tragedy, and even death were often believed to have been caused by conjurers who knew how to manipulate supernatural forces. They employed elaborate rituals, ecstatic dancing, hypnotic trances, concentrated prayer, and fetish objects (gris-gris or juju) hidden near their victim. Birds were common elements involved in both the ceremonies and the fetishes.

For example, charmed objects secretly hidden inside people's beds or pillows "were made of soft, highly colored feathers, brilliant and gaudy, scarlet and gold, bright blue and vivid green, and were about the size and shape of an orange." A nineteenth-century woman in a black New Orleans neighborhood "found some chicken feathers in her pillow very carefully shaped up with fine twine to resemble a rooster. She had a constant headache until the charm was removed."

One practitioner had a recipe for making gris-gris that rivals that of the witches in Macbeth:

> Take the wing of a jaybird, the jaw of a squirrel, and the fang of a rattle-snake and burn them to ashes on any red-hot metal. Mix the ashes with a pinch of grave-dust—the grave of the old and wicked has most potency in its earth—moisten with the blood of a pig-eating sow; make into a cake and stick into the cake three feathers of a crowing hen wrapped with hair from the head of one who wishes an enemy tricked. Put the cake into a little bag of new linen or cat-skin. Cat-skin is better than linen, but it must be tied with the ravelings from a shroud, named for the enemy, and then hidden under his house. It will then bring upon him disease, disgrace and sorrow. If a whippoorwill's wing is used instead of a jay's it will bring death.[385]

A voodoo expert told one investigator in the 1940s that another formula for killing a person was to

> take a rooster, chop off his head and feets, and take them to a graveyard. You can keep the body to cook. When you get to the graveyard, put a black candle in the rooster beak, lay it down in front of a tomb, and bury the feets in the ground in the back of the tomb. Light your candle and pray for one hour that this man have bad luck. Then dig up the feets and take them and the head and go bury them in the man's yard. He won't last a week.[386]

Malicious gris-gris unleashed by one's enemies could also be discovered and neutralized with the help of the proper bird. A "frizzly chicken" was considered a "veritable hoodoo watchdog, for with one of these on the premises a person can rest in peace; it will scratch up every trick laid down against its owner, but if you find one scratching about your house it is a sure sign you have been conjured." Having identified the perpetrator,

> Sweep off a clean place in the back yard and on that clear spot leave a rooster and two hens under a clothes basket all night. Next morning collect their droppings and put them into a bottle of water, leaving it in a "careful place" for three days. Then go at night and make three crossmarks on the door of the person whom you know has tricked you. The trick will be turned against him in less than three days.[387]

Birds could also be employed for good magic. One voodoo practitioner said, "If somebody is already crazy and they wants to get back to good sense, I takes a white dove and splits it in half. I sits the crazy man in a chair and puts the two halves of the dove on top of his head and holds 'em there until all the blood run down his face. After all the blood run out I wash his face in a white bowl and throw the bowl in the river wit' three nickels."[388]

It's impossible to imagine Victorian scientists such as Elliott Coues or Thomas Nuttall believing in the harmless proverbs about birds cherished by their uneducated contemporaries, much less crediting esoteric voodoo practices as effective. Yet more than six hundred folk beliefs about birds and hundreds of arcane rituals involving birds, survived into the twentieth century. Today we are tempted to ask, like those early scientists might have: how could people believe such nonsense?

One answer is that they probably didn't believe it all. Between 1979 and 1982, anthropologist Roger Mitchell spent his summers as a hired hand on a multigenerational, Norwegian American farm in northern Wisconsin. He found that the family still knew and repeated many proverbs passed down through the generations, but they didn't necessarily think they were true. Repeating them was often just an affirmation of their identity, a group acknowledgment that they shared the same past. At other times, the strange old sayings served as entertainment, a kind of performance not meant to be a statement about nature at all. Two or three centuries ago, Midwesterners probably regarded some of this folk wisdom about birds the same way.[389]

In other cases, though, people embraced bird superstitions so wholeheartedly that lives were saved or lost. One researcher found many instances where faith in gris-gris led people to hire voodoo priests to perform rituals, after which they were actually cured of their symptoms. He even witnessed the success of obvious fabrications by charlatans that healed illnesses, "illustrating the power of mind over body." One voodoo healer cautioned Robert Tallant, "Don't ever forget the faith part. You gotta believe." For many people, to believe in witches, conjuring, and omens was simply common sense, and like a placebo, it sometimes worked.[390]

Another way that the great mass of uneducated people made sense of life was to tell tales. Stories helped individuals face challenges and communities bind together. Today, Hollywood and corporate marketing teams impose stories on us, but in the millennia before mass media, individuals invented, memorized, shared, and took comfort from tales that explained the world they inhabited. Many were about birds.

The urge to invent stories, like the impulse to label and classify, comes partly from a universal human desire to impose order on a chaotic world. Reframing life's random events into connected narratives with clear beginnings, middles, and ends causes those events to "display the coherence, integrity, fullness, and closure of an image of life that is and can be only imaginary." When researchers left a tape recorder running nonstop in a working-class home in Baltimore in the 1980s, they discovered that ordinary people framed their experience as narratives at an average rate of 8.5 times per hour.[391]

But stories also help us define where we belong. Children figure out how the world works by hearing stories from their elders. "In de early days," a former slave recalled during the Depression, "Ah heerd many an' tale 'bout ole Brothuh Rabbit what woke me to de fac' dat hit tecks dis, dat an' t'othuh to figguh life out—dat you hafto use yo' haid fo mo'n a hat rack, lack ole Brothuh Rabbit do." People embed themselves within their culture by recognizing and retelling the canonical tales, the stories that everyone knows. Traditional storytelling was an important way that people knew they were Yankee or Yankton, Creole or Cree.[392]

We know that the Ho-Chunk and other Midwestern tribes have told and retold the Red Horn cycle for a thousand years. How does a tale remain intact through all those retellings? According to former Ho-Chunk tribal president Chloris Lowe Jr., participants sit in a circle and the first person recites a traditional narrative. The next person recites it again, and if he makes a mistake he must begin the recitation all over again. Through multiple sessions with multiple listeners checking for accuracy, stories like those about Red Horn could maintain their integrity through the generations. Whether a tale was a sacred creation story or a traditional secular tale, it explained some aspect of how the world works.[393]

A good example is this Plains Indian story explaining why turkey vultures have no feathers on their heads. Its hero is Ishjinki, a semidivine shape-shifter and trickster of the Missouri, Otoe, and Iowa tribes:

> As Ishjinki was travelling, he came to a place where he saw Buzzard flying above him. "Oh, grandfather," he said, "how much fun you must have up there in the air. There is nothing that can hurt you, and you can see everywhere. I wish I could get up as high as that and see as far as you do."—"You would never get used to it, my grandson," said Buzzard. "You belong down there and I belong up here. I'd rather you stayed where you are."
>
> But Ishjinki begged and teased Buzzard, until finally the bird took him up a little ways and returned. Then Ishjinki asked Buzzard to take him higher. This happened four times. The last time Buzzard took Ishjinki very high, so that Ishjinki cried "Wahaha" every time that Buzzard dipped and soared. Finally Buzzard went down close over the tops of the timber until he saw a hollow stump. He tipped his wings, and threw Ishjinki headfirst into the stump. Ishjinki was stuck there.

After Ishjinki is rescued by friendly Indian women, the story continues:

> Ishjinki was very angry at Buzzard, so he made a trap to catch him. He pretended he was a dead horse and lay still until the crows picked at his rump. Buzzard appeared, but even though Ishjinki tried his trick three times he could not fool him. The fourth time he tried being a dead elk. The birds came and ate most of his rump. The crows even went in and out of his body. At last Buzzard came and pecked at Ishjinki's rump. Finally, he stretched his head and reached way inside. All at once Ishjinki closed the opening. "Now I've got you," he exclaimed, and he walked off with Buzzard dangling from his rear. He kept Buzzard there for a long time. Finally he said, "I'll let you go now. You have paid for your trick." Buzzard then pulled his head out, but he has been bald and smelly ever since.[394]

French immigrants in the Mississippi Valley had their own creation tales, too, like this one that begins with Noah sending the raven out to look for land after the biblical flood:

> He opened a window and let out a raven, which until then had been the bird favoured from creation with bright and magnificent feathers and whose voice filled the air with joyful warblings. 'Fly over the surface of the earth,' Noah commanded the raven. 'If you find there any green trees, bring me a branch.' The raven saw some bodies floating in the water and, satisfying his voracious appetite, set out to devour them. He forgot his master's order, and no longer returned to the ark which had run aground on a mountain. Noah then cursed the raven for its faithlessness; his curse blackened the feathers of that bird and changed its warblings into a hoarse and plaintive croak . . .

One collector heard an old woman recount this legend of the crows to her children: "She saw, with satisfaction, in their eyes the impression that her tale made on her small audience. She finished with the words: 'If God's anger falls like that on the simple birds who were cursed by Noah, how much more his punishments will weigh on the faithless man who disobeys his commandments.'"[395]

Here is another didactic tale that was carried north by African Americans into Michigan during the Great Migration:

> Mr. Crane called to Mr. Buzzard, "Why don't you build a house, Mr. Buzzard?" He said, "What do a man need with a house? Look how nice and cool this wind is blowing." Buzzard was sitting in the shade, wings apart to let the air through, just resting easy. That morning early it begin to thunder, lightning, and rain. Mr. Buzzard begin to get wet. "Well the next sunshiny day I'm sure going to build me a house." So the next morning the sun was bright and pretty and the wind was blowing like it did the day before. Said "Shaw, what do a man need with a house now?"[396]

Similar values were taught in this African American tale of a debate between a lark and a partridge, collected in Missouri in the 1880s and rendered in dialect by Martha Young:

> When Fiel' Lark and Partridge was young, dey had a talk, but dey could n't agree. Nairy one would let loose his end of de argument. Bof b'lieve and 'clar' de b'lief dat dey knowed bes'.
>
> Fiel' Lark say he gwine be up, up, up while de sun's low, and he say dat'll insu' him a long life. He so do and he's a long-livin' 'dustrious bird.
>
> Partridge say what's de use er livin' long ef you gotter all de time work hard. He say better run low in de grass and live by rich pickin's. He say better be stout and de worl' don't know you, dan slim and de sun show you.
>
> Fiel' Lark, he say:
>
> "Ef you do dat
You'll git too fat.
And Mister Man'll kill you right now!
Des now!"
>
> Partridge, he make ansah:
>
> "No fear!
Don't keer!
Good cheer!
While I am heah!
Don't keer!"
>
> Dem two keep dat up all summer long. All de whole endurin' time, dem two quar'l des dat way. Up go de Fiel' Lark whilst de sun's yit low:
>
> "Dis laziness'll kill you!
Dis laziness'll kill you!
Des now!
Right now!"
>
> Partridge runnin' and hidin' and scootin' low in de bushes, he whistle clear:
>
> "Don't keer!
Don't keer!"
>
> He don't keer bit neither. Ever' year he round and fat. He git kilt up, too. But twel he do, he keep whistlin':

"Don't keer!
Do you hear?
I don't keer!"

Fiel' Lark, he fly so high, he work so hard, he eat so little dat his wings is thin, his legs is slim, his body so po' nobody want ter eat him. He de long liver. He's up wid de day, singin' all de time:

"Dat laziness'll kill you!
Right now!
Des now!"[397]

The French and British colonists who arrived in the Midwest during the seventeenth and eighteenth centuries brought traditional tales from their homelands. Fully half of the tales collected from English speakers by folklorists could be traced across the Atlantic. Stories that came down into the nineteenth century included many retellings, with some alterations, of immediately recognizable European tales about dragons, princesses, giants, and so forth. Among the most popular French tales were supernatural stories about spirits and ghosts whose Breton and Norman seeds bore rich fruit in the western wilderness.[398]

Folklorist Richard Dorson collected this ghost story about an owl in St. Ignace, Michigan, a small town in the Upper Peninsula where Lake Huron, Lake Michigan, and Lake Superior come together. It was told by "Aunt Jane," a working-class woman born about 1860 who framed it as an authentic recollection:

Old Sarah [Champain] was very sick all the time, she was choking. Every night she took ill and had to go to bed, when a great big *ghibou* [owl] came and perched on her clothes line. It came the same hour every night. It is unusual, you know, to see an owl right in the City. Mrs. Champain got the doctor and the priest but they couldn't tell her what was wrong.

The clothes line ran right out under the window—she was in an upstairs apartment—right next to my aunt's house. The door of my aunt's room was open in the summer, and she would see the owl when she went to take the clothes in. She told her nephew to kill the *ghibou* because he would get the chickens. So he takes a shotgun and shoots the *ghibou*. He sees it flop over, puts the gun away, and goes to pick up the *ghibou*, to cut it up for a feather duster. But he can't find it. He goes round the house to the highway, down to the trail—the building was set up on the hill—and sees old Mrs. Lozon, lame, trying to crawl up the hill. She'd been perfectly all right the day before, but now she could hardly walk. Charlie Lottie picked her up in a rig and took her over to one of her sisters. She never got over it—she stayed there 'till she died.

The man, Jimmy Vallier, ran back, all excited and numb, crying, "That was no *ghibou*, that was a *roup-garou* [*loup-garou*, or shape-shifter]!" Jim said he was afraid to follow after Mrs. Lozon because she might do something to him, to keep her secret. My aunt laughed and said, "That was only a *ghibou*; I saw it right out the door." My aunt changed her mind later, though. And do you know, when Mrs. Lozon died, old Sarah got better right away.[399]

English speakers, like the French, brought classic legends from their homelands, but they also invented new tales about local characters and events, especially humorous ones. Among British settlers, jokes and anecdotes set in local places outnumbered traditional longer narrative forms. Yankees developed particular skills at telling tall tales, stories of exaggerated proportions told with apparent sincerity that were intended to test the listener's gullibility. Beginning in eighteenth-century New England, if not before, many tall tales revolved around hunting, especially the remarkable shot. "Hunters who cocked a long-rifle in the day time," wrote Dorson, "habitually drew the long bow of evenings."[400]

As Yankees moved west into the continent's heartland during the late eighteenth and early nineteenth centuries, they spun many tales like this one collected in Wisconsin:

Ben was very fond of hunting and was a crack shot with a rifle. It is said of him that he often let his farm work take care of itself while he went into the woods for game. Once, when hunting, he saw about two dozen pigeons sitting on the long limb of a big tree. Ben wanted those pigeons, but he had only his rifle which fired a single ball. Then he got an idea. He moved carefully until he was opposite the end of the

> limb. Taking careful aim he fired. His ball split the branch, the legs of the perched pigeons dropped into the crack which closed on them and the birds were trapped. Ben then cut the limb from the tree with another shot or two. He went home carrying the limb over one shoulder with the pigeons dangling from it.

This motif—the hunter splitting the branch to catch many birds at the same time—was found in dozens of stories collected from New York, all across the Midwest, south to Texas, and as far west as Alberta.[401]

Tall tales often concerned birds, such as this one, from a man who couldn't believe that passenger pigeons had been exterminated by human actions:

> "It cain't be," one old fellow told me, "there was too many of 'em. If every man, woman, an' child in this country had et pigeons every day, an' nothin' but pigeons, it wouldn't have made a dent in them big flocks." . . . most of the old-timers are somehow persuaded that they all perished in the ocean. Not only this, but the hillfolk believe that many ships lost at sea in the 1880's and 1890's were really destroyed by pigeons. The story is that the pigeons tried to cross the ocean, but became exhausted. Whenever they saw a ship the weary birds alighted on it in such numbers that their weight actually sank the vessel. If the crew took to the boats, the pigeons sank them, too. I heard an old man at Fort Smith, Arkansas, describe such a shipwreck in a way to make one's blood run cold. Several hours later, it occurred to me that the old man had never seen the sea or any boat bigger than an Arkansas river-packet.[402]

The tall tale may have peaked in the 1880s and 1890s with oral stories about the mythical lumberjack, Paul Bunyan. Although Bunyan was later co-opted by marketers and advertisers, an oral tradition of stories told aloud by loggers was collected between 1900 and 1920 that predates the commercial "fakelore." Birds sometime feature prominently in these tales about Bunyan.

Bunyan was a terrific hunter and "could shoot ducks and geese flying so high in the air that they were often spoiled before they reached the earth.

So he salted his shot to preserve them." Unsuspecting listeners were told that the remote forests around his logging camps were populated by birds unknown to science: "The hillside plover nested on the slopes of Bunyan's famous Pyramid Forty. Living in such a locality it laid square eggs so that they could not roll down the steep incline."[403]

When I first read these traditional tales, proverbs, and maxims about birds, they seemed quaint at best; more often, silly or simply bizarre. It's difficult to imagine that such fabulous notions were ever taken seriously by anyone. The mental habits of science—observing, classifying, speculating, testing, judging, concluding—have permeated our culture so thoroughly that there's no room for folklore except as a childish game or a charming eccentricity.

But earlier generations hadn't internalized the scientific worldview, and folklore was more meaningful to them than it is to us today. For many people, these beliefs explained incomprehensible features of the natural world. For others, folklore may have reflected what they simply wanted to be true. We interact with the world not just through our understanding, but with our desires, too. Even if our ancestors didn't literally believe everything they said about birds, their proverbs and superstitions reveal their hopes, fears, and values.

10

The Great Extermination

When I told my wife's uncle that I was researching how earlier people interacted with birds, he stretched one arm in front of the other, sighted across his wrist, pulled an imaginary trigger, and said, "Blam!" He was right, of course, but he didn't appreciate the true significance of his gesture.

For twelve thousand years, humans and birds lived in relative harmony in the heart of America. People harvested birds in sufficient numbers to augment their diet, make tools and clothing, create art, and embody spiritual wisdom. Populations of both birds and people waxed and waned over the centuries in different places, but the ancient balance remained steady until the nineteenth century, when new technology opened a Pandora's box of uncontrollable forces.

Then, in a single human lifetime, bird species that had flourished in the Midwest for millennia were wiped from the face of the earth. Others narrowly escaped extinction but were decimated as thoroughly as the bison massacred on the Great Plains and the ten-story white pines clear-cut from northern forests.

Before the mid-nineteenth century, humans didn't have the power to affect bird populations in such a devastating way. Tribal communities were comparatively small, and their traditional hunting techniques—atlatls, nets, traps, bows and arrows—could not wipe out an entire bird species. The numbers of early European settlers and the hunting tools they brought across the Atlantic were also limited, but in the mid-nineteenth century Americans adopted technology that had a devastating effect.

French explorers arriving in the Midwest in the seventeenth century, and the British who followed them in the eighteenth, carried guns that were more like miniature cannons than today's rifles. They averaged four to five feet in length and weighed as much as nineteen pounds. To fire one, the hunter wrapped a tablespoon of gunpowder in a piece of paper or cloth, tucked it in the muzzle of the barrel, stuffed in a half-inch ball of lead or a handful of shrapnel, and rammed all this down to the far end of the barrel with a rod.

The hunter then stretched a spring tight by cocking the gun's three-inch hammer, sprinkled a little fresh powder into a small pan next to the near end of the barrel, rested the gun on some stable object like a tree limb, pulled the wooden stock firmly against his shoulder, and pulled the trigger. The spring unlatched, the hammer scraped a piece of flint across a ragged steel plate, a shower of sparks fell into the small pan of gunpowder, and an explosion shot the ball out the barrel.

These guns were called flintlocks, and they often required several attempts before actually firing. Damp powder, worn flint, a dirty barrel, or a ball that was not truly round could each undermine a successful shot. If the hunter missed, reloading and firing a second time took anywhere from three to ten minutes, by which time the bird was long gone. Just a few shots could be taken before the barrel and firing mechanism had to be disassembled and thoroughly cleaned.[404]

Many models of flintlock were made in Europe and America between 1650 and 1850; generally, the greater the barrel length, the more accurate the shot, and grooved barrels further improved accuracy. Some late-eighteenth- and early-nineteenth-century Kentucky rifles had five-foot grooved barrels enabling hunters to reliably hit targets at a distance of fifty to one hundred yards and could be reloaded in as little as thirty seconds. Various versions of the flintlock musket or rifle were carried west by millions of settlers who flowed down the Ohio into the Mississippi Valley or sailed up the Great Lakes and onto the Plains before the Civil War. This was the technology on which countless American families depended for food and protection for 250 years.

"Hangin' over the fireplace, or the door, was purty sure to be a flintlock rifle," recalled Oliver Johnson of his childhood in Indianapolis during the 1820s and 1830s.

> It was mighty essential to furnishin' the family with meat. Most every man was handy with a rifle, and some women wasn't so bad either. When a boy was big enough to carry a rifle and knew how to load it, he was allowed to hunt with it. He had to bring in game and kill squirrels by shootin' them in the head, or he was told to leave the rifle alone. Old timers didn't believe in wastin' powder and lead.[405]

Johnson's account of precisely how hunters used their firearms is worth quoting in full, since the step-by-step process he describes was once as familiar to most Americans as sending a text message is today:

> Besides the rifle, the equipment of a hunter was a shot pouch and powder horn. The pouch, about six or eight inches square, made from deer hide with the hair left on, was carried on the right side by a strap runnin' over the left shoulder. In this pouch you carried your bullets and patchin'. The powder horn was fastened to the same strap and hung over the shot pouch. Hunters took lots of pride in their powder horn. It was made from a cow or ox horn, scraped and polished until it was so thin and clear you could see the powder through it. The small end was fitted with a wood plug, the big end with a block of wood. Up above the powder horn and fastened to the shoulder strap in front was a small scabbard riveted together with lead rivets so as not to dull the edge of the small butcher knife carried in it. Hangin' by a short leather thong was the powder charger. It was a measurin' cup, made from different things; Pap's was the tip of a deer horn . . .
>
> The first thing I learned was the proper procedure in loadin'. The rifle was set on the ground with the barrel held by the left hand, the charger grasped between the forefinger and thumb of the same hand. With the right hand, pick up the powder horn and pull the plug with your teeth. Pour the powder in the charger and replace the plug. Drop the powder horn and pour the charge of powder down the barrel. Then get the tallered linen patchin' out of the shot pouch, put it over the end of the barrel, and push a bullet down on it with your thumb as far as it will go. Draw your butcher knife and cut off the patchin' across the end of the barrel. Then push the bullet down against the powder with the ramrod. With the old flintlock rifle there was no cap

> to bother with. Enough powder would dribble out in the pan to flash when the flint made a spark. If the flint was in bad shape, or the powder in the pan got damp, she wouldn't go off. Occasionally I got into trouble with mother when she found my shirt tail full of holes, after I had run out of patches and used it as a substitute.[406]

Given the unreliability of their weapons and the importance of birds as food, subsistence hunters like Johnson sought species that were large and easy to approach, and yielded a substantial amount of edible flesh, such as turkeys and waterfowl. "A favorite way of huntin' turkeys," he recalled,

> was with a caller. We made it from the flat bone of the second joint of a turkey wing. It took experience to make "turkey talk" on one of these callers. If you made the wrong sound, the turkeys knew the difference right away and was gone in short order. When you found fresh signs of turkeys in the woods, you got behind a big log and hid yourself. Then by a peculiar suckin' sound made on the caller with your mouth, you give three "keouks," imitatin' a turkey. If there was an old gobbler in hearin' distance, you'd get three "keouks" for an answer. You call again. After gettin' a answer or two, the gobbler would run, with head low, for a short distance, thinkin' he was goin' to join up with another flock. All his flock would come trailin' after him. They would stop, straighten up their heads and wait for another call. Then they would answer and run toward you again. When they was almost near enough for a shot, you lay low and put your mouth near to the ground and give the last call. That made it sound farther away. Then you raise your head and shoulders above the log, take quick aim and fire.[407]

Turkeys were favorite game birds because, when treed, they usually stayed put even after the hunter started firing, until many or even all of them had been killed. In 1791, British traveler Patrick Campbell encountered a soldier who, while held captive by the Iroquois,

> was one day permitted to go along with them to the woods on a shooting party; that soon they fell in with Turkies, the Indians pursued on foot as fast as they could run, bawling and hallowing all the

> time to frighten the birds, and when they had thus got them upon trees, that they shot many of them . . . Several other persons told me that was the surest way to get them. They are so tame or stupid when they are in the trees, as to stand perhaps till the last of them be killed; whereas, on the ground, they are so quick sighted and fleet, that in an instant they are out of sight.[408]

Ruffed grouse exhibited the same suicidal tendency as turkeys. Hunting on the shore of Lake Erie in October 1796, Isaac Weld reported, "It seems that you must always begin by shooting the bird that sits lowest on the tree, and so proceed upwards, in which case the survivors are not at all alarmed. Ignorant, however, of this secret, I shot at one of the uppermost birds, and the disturbance that he made in falling through the branches on which the others were perched put the flock to flight immediately."[409]

Grouse and other game birds, though smaller than turkeys, were an important food source for early settlers. "In travelling through these Prairies," a visiting Englishman wrote from southeastern Illinois during the winter of 1822,

> every one must be struck with the vast number of a species of grouse, called "Prairie Fowls" [Greater Prairie Chicken] . . . they are delicious eating, and are killed in great numbers by the unrivalled marksmen of this country. After driving up a flock of these birds, the hunter advances within fifteen or twenty paces, raises his long heavy rifle, and rarely misses striking the bird on the head. I have witnessed over and over again this surprising accuracy, and have fired away numberless pounds of lead in trying to imitate it, but without success. I contented myself therefore with shooting the birds in the body, by which I rather tore and spoilt them. But, however difficult I found it to hit a bird anywhere with a single ball, the Backwoodsmen regarded my unsportsmanlike shooting with as much contempt, as one of our country squires feels, when a cockney shoots at a covey of partridges on the ground.[410]

Waterfowl also sustained early settlers in the Midwest. Their numbers and delicious flavor made them a favorite target. "I have already remarked in my Journal," wrote Jonathan Carver in about 1766, "that the teal found on the Fox river, and the head branches of the Mississippi, are perhaps not to be equaled for the fatness and delicacy of their flesh by any other in the world." Farther south, in April 1816, Timothy Flint found "innumerable multitudes and varieties of water-fowl, of different forms, and plumage, and hues, were pattering in the water among this grass; or were raising their several cries, as we frightened them from their retreat. We easily obtained as many as we wished; and when roused to the wing by our guns, they soon settled down in another place."[411]

Canada geese often left wetlands to come into farmers' fields where they were hunted as a source of food and warmth. Rose Taylor of Dane County, Wisconsin, recalled that around 1870,

> As the odd girl in the family, I was permitted now and then to go on a hunt with the boys. If on a goose hunt, we were in the cornfields long before the eastern sun colored the dull gray clouds resting on the horizon. We hid inside the corn shock with the old musket sticking out and ready to pour its charge into the first flock that came. They usually lighted cautiously in the field at dawn and began feeding. At a favorable moment, one of us pulled the trigger and often a Canada goose or two lay lifeless in the field. It was a thrill to go home with them. The breast of the bird is solidly packed with the finest feathers and closest to the breast is the silky down. These goose feathers were used for pillows and for feather beds, the latter being especially needed during the extremely cold midwestern winters.[412]

Settlers on the Midwestern frontier were far removed from technology's cutting edge, so when European travelers passed through equipped with the latest rifles or double-barreled guns that could shoot birds in flight, their American hosts were very impressed. For example, British visitor William Blane noted that in Illinois in 1822, "Shooting flying is an art wholly unknown to the Backwoodsmen. Indeed I have often been amused, when speaking to them upon this subject, to see with what scepticism they have received my accounts, gravely asking me, whether I really meant that any one with a double-barrelled gun could kill two birds on the wing, one after another."[413]

Although the new shotguns were available in the east by 1850, most settlers in the Midwest still used antiquated flintlocks, which were slow and inaccurate and had little impact on bird populations. "Guns were few and of poor quality," wrote pioneer Clay Merritt of central Illinois in the 1850s. "The farmers and householders had only what they brought with them when as pioneers they came into the country. Rifles prevailed; muzzle shotguns were few and hardly obtainable . . . What birds were killed were perforated with rifle balls and of little value."[414]

Nearly every visitor to the Midwest wrote about the astounding numbers of birds encountered before the invention of shotguns, as did the pioneer settlers when they looked back in old age. The abundance of game had been a common theme in accounts of the continent's heartland since its earliest exploration. "Of all the nations of Canada," wrote Father Sebastien Rasles in 1689, "there are none who live in so great an abundance of everything as the Illinois. Their rivers are covered with swans, bustards, ducks, and teals. One can scarcely travel a league without finding a prodigious multitude of turkeys, who keep together in flocks, often to the number of two hundred." His colleague Father Jean St. Cosme wrote from Illinois in the winter of 1699, "One cannot fast on this river, so abundant is it in game of all kinds, swans, geese, ducks."[415]

As long as white settlers were few and far between, birds continued to reproduce as they always had. Toward the end of the eighteenth century, fur trader Peter Pond could still write from the Upper Fox River in central Wisconsin that "the Wild Ducks When thay Ris Made a Nois like thunder. We Got as meney of them as we Chose fat and Good . . . You Can Purchis them Verey Cheape at the Rate of two Pens Per pese. If you Parfer shuting them yourself you may Kill what you Plese."[416]

But the vast quantities of birds retreated in the face of encroaching humans. Settlers streaming down the Ohio and up the Great Lakes at the turn of the nineteenth century drove the seemingly limitless flocks off the main thoroughfares and into remote, less-traveled locations.

For example, when Victor Collot had gone down the Ohio River in the summer of 1794, he reported that near Louisville, "Both sides of the river were covered with game, chiefly waterfowl, in such quantities that it seemed scarcely possible to augment the number: geese, ducks, swans, herons, and roebucks [deer], were mingled together, and line both sides of the Ohio." Yet just fifteen years later, Fortescue Cuming traveled the same route and found, "Travellers descending the river have but little chance of obtaining any game, as its having become so great a thoroughfare, has rendered both the four footed, and feathered tribes fit for the table so wild, that it is rare that any of them, even when seen, can be shot."[417]

The same thing happened along the shores of the Great Lakes. In 1812, a visitor noted that "geese are not plenty in the waters of Lakes Ontario and Erie at present, but used to be before the country was settled by white people, yet they are plenty enough in all the lakes north of the settlements." In June 1823, Thomas Say found at Chicago that "although the quantity of game in this part of the country is diminishing very rapidly, and although it is barely sufficient for the support of the Indians, still there is enough, and particularly of the smaller kind, to offer occupation to the amateur sportsman."[418]

Far from the major settlements and principal waterways, bird populations remained stable for another generation. While numbers were declining on the Ohio River and at Chicago, in rural Michigan, "Wild geese and ducks [were] found in such immense flocks on the lakes, rivers and bays, that their vociferous squalling and the thundering noise of their wings, seem[ed] to remove all apprehension of the fear of man." As late as 1860, in central Illinois, Clay Merritt reported: "On a cold day in early winter I have seen acres covered with the birds [prairie chickens], as thickly as they could sit."[419]

This was because in rural areas settlers possessed crude flintlocks and generally hunted no more birds than they could eat. Henry Herbert remarked that in the 1840s, "The country farmers, as a body, have neither the time, the inclination, nor the opportunities for making themselves acquainted with the names, habits, or manners of game-animals." Although they all took geese or turkeys for the table, when he showed them quail and woodcock he had killed, he was often "met by a remark that the speaker had lived on that farm all his life, and had not seen a dozen such birds in his life-time—and the name of the bird was unknown to them."[420]

But starting in about 1840, technological and demographic forces converged to launch one of the greatest slaughters of one animal by another in the history of the planet.

Between 1830 and 1870, millions of Americans moved from self-sufficient family farms into congested cities. No longer able to raise their own food,

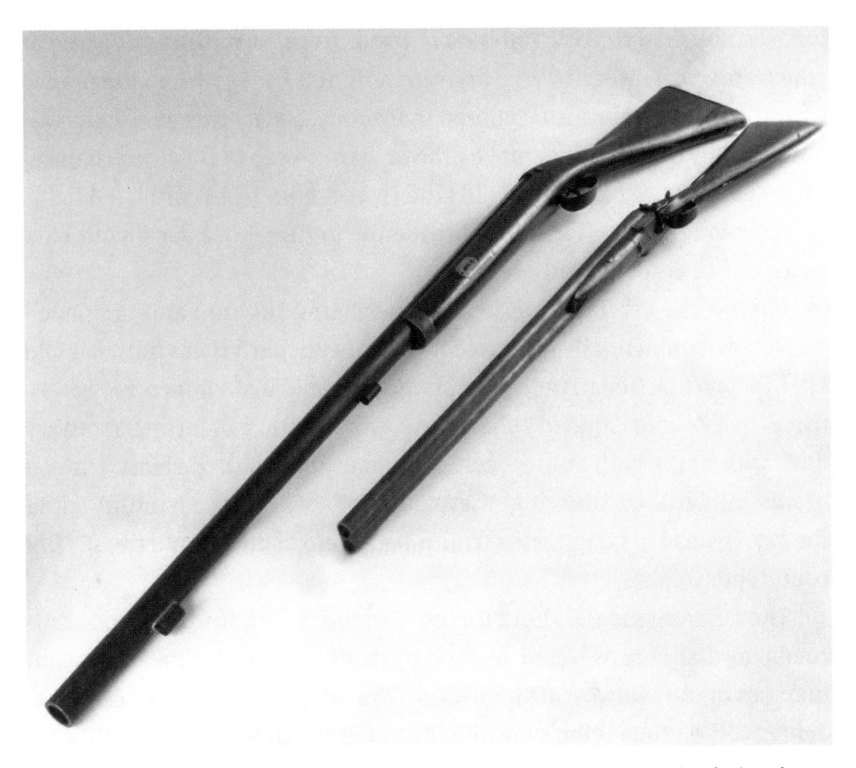

The twelve-gauge shotgun on the right was made in Madison, Wisconsin, during the 1870s. The massive "punt gun" beside it has a one-inch diameter barrel and weighs twenty-six pounds. In 1876, a hunter killed ninety-six ducks on Horicon Marsh with a single shot from a similar weapon. WISCONSIN HISTORICAL MUSEUM OBJECTS 1947.1743 AND 1979.96.2

they had to be fed by commercial markets eager to obtain provisions. The rapid spread of the railroad throughout the Midwest during the 1850s enabled suppliers to deliver not only wheat and corn but also geese, ducks, grouse, and other game birds to these urban markets for the first time. A prairie chicken shot in a Missouri field at dawn might be served in a St. Louis restaurant the same night, or, after being packed in sawdust and ice, sold in a New York market three days later. Businessmen from Chicago who hunted for sport could board a northbound train in the city, spend a weekend shooting in Wisconsin's Horicon Marsh, and be back at their desks Monday morning.

During the same decades, waves of new immigrants flooded onto the prairies and plains, where they uprooted native plants and drained long-established nesting habitats in order to plant millions of acres of wheat and corn. Merritt, who arrived in Illinois in 1857 when the prairie was still largely unbroken, claimed that Midwestern birds were exterminated not by overhunting but "by the onward sweep of those mechanical forces which have converted wild lands into smiling and cultivated fields . . . the plough of the farmer has made the prairie, once vocal with wild birds, only a desolate solitude."[421]

Finally, the invention of the breech-loading shotgun and preloaded cartridges exponentially increased the number of birds that a hunter could kill in a day. Starting in the 1850s, shotguns could instantaneously send a three- to five-foot cloud of pellets forty yards. Instead of hitting a solitary duck with a rifle ball, hunters could kill many with a single blast. Instead of missing birds on the wing, they could reliably knock a handful out of the sky. Instead of taking several minutes to reload, they could fire several rounds per minute.[422]

The convergence of these forces—urbanization, immigration, railroads, and shotguns—sent legions of hunters into the meadows and marshes of the Midwest after the Civil War, where they took an unprecedented toll on the region's birdlife. Although the first licensing and conservation laws had been passed, they were not effective, so hunters killed as many birds as the market demanded—and the market was insatiable. Within two generations, the clouds of passenger pigeons that had darkened the skies for centuries and the rafts of waterfowl that had crowded lakes so that canoes could hardly pass had vanished forever.

The passenger pigeon has become America's best-known example of extinction. Much has been written about flocks that blocked the sun, entire towns dropping everything to hunt them, railroad cars stuffed full with their carcasses, and the last pigeon dying alone in 1914 at the Cincinnati Zoo. Almost as famous is the story of the assault on marsh birds and songbirds in the late nineteenth century, the feathers of which were displayed on hats. Every birder has read about the "plume hunters," the first game wardens who fought them, and the birth of the Audubon Society. The role played by habitat loss is also well understood; bird populations were decimated as Midwesterners blanketed the landscape with farms, drained its wetlands, and replaced native grasses with cultivated grains. But the history and consequences of commercial bird hunting are not as well-known, even though market hunting played an equally large part in the extermination of American birds.

Market gunners, as they were called, bear some similarity to lumberjacks and cowboys, two other vocations that flourished in the same decades and attracted rugged outdoorsmen. The major difference between them was that market gunners saw their work as sport rather than labor. "Hunting was looked upon as so much of a pleasure," recalled Clay Merritt, an Illinois game broker who hired many market hunters for half a century, "that the immense toil and labor endured was not counted." It attracted men who would rather shoot than work at steady indoor occupations, men who tended to drink heavily, live recklessly, and give little thought to the future. Merritt recalled:

> With two or three notable exceptions, I cannot find that their lives have been sweet or savory, or that they have served any high or beneficent purpose upon their contemporaries . . . For many years hunting was more profitable than farm work, and was as good as a trade. Money was made easy and spent freely. Such men pair their vices with their virtues, and hope to balance the account, but vices are negatives, they start at zero and go downward . . . Such men float easily along on the stream of good luck, as they call it, when it passes their way, and become too exalted over success. With the ebb tide they fall back into the trough of their many wasting burdens again.

He regretted that many hunters he had known lost their way in "the saloons where morality is constricted, and disease blossoms and sheds its baleful fruit." Yet in the end he concluded, "There was always a brighter side which was more or less obscured. I know of very few of them, but what if they had chosen could have reached higher walks of life, persistently and deliberately attempted."[423]

Dick Harker of northwestern Iowa was probably a typical market hunter. He gave a long interview in 1945 that reveals exactly how most of the waterfowl and game birds of the Midwest were wiped out during the late nineteenth century.

About 1887, Harker was hired as a hunter by the firm of Winter Brothers of Spirit Lake, Iowa, which sent him across the state line to Heron Lake

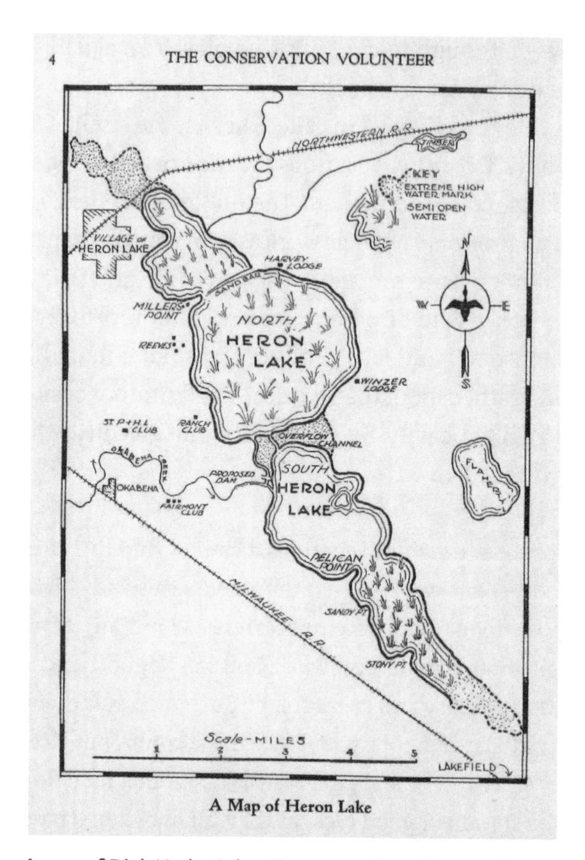

A map of Dick Harker's hunting grounds at Heron Lake in the *Conservation Volunteer*, January 1941. *MINNESOTA CONSERVATION VOLUNTEER*

in Lakefield, Minnesota, to collect ducks for delivery to urban markets. "We built boats out of ordinary lumber," he said, "peaked at both ends so we could push them through the rushes. We couldn't row in the lake, there were too many rushes. We used push poles." Harker and six other hunters spent several weeks there shooting ducks. "We all kept a record of how many we got," he recalled.

> When night came we asked how many Redheads, how many Mallards, and so forth, we killed, and we kept a book account of every bird we killed. I have a brother that still has his records. I used to kill about two thousand birds from the fifteenth of August until the time

it froze up. This is an estimate of what we used to get when we were hunting for Winter [his broker]:

Dude Gilbert 3,000
Joe Winter 2,000
Fred Winter 2,000
Jim Flaherty 2,000
Dick Harker 2,000
Russ Klein 1,500
Cornell 1,500
[Total for a single autumn]: 14,000.[424]

Then, in the spring, they shot upland game birds: "We killed more Golden Plover than anything else. In about one month, one spring, I killed two thousand Golden Plover, Grass Plover, Curlew, Jacksnipes and Yellowlegs."

They carried shotguns and filled their shells by hand: "We used number nine and ten shot to kill plover. We loaded our own shells. Generally, it wouldn't cost us over a half-cent for the light load and a cent for the heavy load. We used brass shells almost altogether. The last year I hunted for the market [ca. 1904] we began to buy the loaded shells with black powder."[425]

At first, Harker, like most market gunners, worked for wages: "Winter [the broker] paid us seventy-five dollars a month and furnished us with everything: food, shelter, shells, everything except our guns. This was for both spring and fall shooting." But he had a realization when "one week, I killed enough Redheads and Canvasbacks to pay my salary for the whole month." So he and his brothers struck out on their own in about 1890: "There were three of us—two brothers and myself. We were all good for about three thousand snipe and plover. We killed them in about one month . . . My brothers and I shot for the market for fourteen years [ca. 1888–1902]."[426]

Harker's interview, like countless nineteenth-century magazine articles and recollections of old hunters, overflows with tales of big days, rare kills, and extraordinary shots. This one can represent the entire genre:

At one time, up in Minnesota, when we were hunting for the market, my brother, Frank, and I came down on what they call Grass

> lake and about a quarter of a mile ahead of us we saw something that looked peculiar to us; it was spotted. The more we looked the more we thought it was funny, and then we saw the wings flutter and it was nothing but Green Wing Teal and there were possibly three or four thousand. The wind was just blowing right to the shore and it was the best place in the world to get them. We had to shoot through the flock. We both had double barreled Smith guns. I said to my brother, "Are you ready?" and he said, "Yes," and I said, "Let's give it to them," and we raised up and I didn't hear his gun go off, and then it finally went off. He had had his gun on safe. He got a few. I killed thirty-seven. At Heron lake we would sneak around and shoot into the ducks when they were sitting in the water and would kill eight or ten at a time. I would kill three or four sitting and the rest when they got up.[427]

To store their kill, Harker and his brothers built a freezer modeled on a common icehouse.

> You build a fourteen foot square in the center out of tight lumber. We put on a two by six border and filled the space with sawdust. Then we put on another two by six, left that as a dead air space, and then put on another two by six and filled that with saw dust, giving us a wall eighteen inches thick. Then, there were about twenty galvanized pipes filled with packed ice . . . We had two troughs to drain the water away. We had shelves built about eight inches apart and we could freeze a duck in there in one night as hard as a bullet.[428]

When they had enough birds for a shipment, they would send them to the city tightly packed in ice and sawdust. "We shipped our birds to Chicago and New York," Harker explained. "Sometimes they would come out to buy them, but most of the time we shipped them out by express; and express was high in those days. We shipped them to commission houses."[429]

The best-tasting birds were always the most desired. Although demand was steady, supply depended on seasonal migrations, which caused prices to fluctuate. Harker recalled that in the fall, "We used to get a dollar and a half a dozen for Saw Bills, Blue Bills, Whistle Wings, Butterballs and such

This unidentified late-nineteenth-century photograph shows a pair of hunters with their dog, guns, and more than twenty waterfowl. WHI IMAGE ID 59473

like of that [mergansers, ruddy ducks, goldeneyes, and buffleheads] . . . in the spring of the year we got ten cents a piece for them, that is for a mixed bunch of ducks . . . We used to get thirty-six dollars a dozen for Canvas-backs. This must have been along in the nineties. The ducks would have to weigh so much to bring that price."[430]

According to the 1890 US Census, the Harker brothers were among 2,534 people who identified their occupation as "hunters, trappers, guides [or] scouts." Nearly 40 percent of them worked in the Midwest, Ohio Valley, or eastern Plains; a third were illiterate, and a quarter had been unemployed at some point during the previous year. In the Midwest, the largest number worked in Michigan and Wisconsin, each credited with more than one hundred professional hunters, followed by Minnesota, Illinois, and Missouri, which employed several dozen each.[431]

If market gunners averaged two thousand birds per season as Harker did, and hunted in both spring and fall, then Midwestern hunters were killing about 2.8 million game birds and waterfowl annually. Nationwide, the total would have been more than ten million birds sent to market each year. The true total was, of course, even larger since those numbers ignore the birds killed by hunters who didn't identify themselves in the census as professionals and shot for the market only part-time each spring and fall.

Unlike the Harker brothers, most market hunters didn't get involved in shipping birds to urban dealers. Instead, they sold their game wholesale to a local broker like Clay Merritt, who handled waterfowl and game birds in the Upper Mississippi Valley from 1858 until 1904. After half a century in the trade, he published a delightfully eccentric memoir that rings with a Midwestern twang reminiscent of Mark Twain. Merritt describes not only the mechanics of the market hunting business, but also the beliefs, desires, and values of the people who composed it.

Merritt moved from New York to Illinois in 1855, where he tried his hand at farming and teaching and got married. Two years later, he headed home with his new bride for a visit to their families. Before boarding the train in Illinois, as an experiment he shot "quite a number" of plover and prairie chicken. "These I put inside the trunk," he recalled,

> where, after a passage of three days, it was no surprise that the birds I had were a little tender, but I took them out of my trunk and proceeded to sell them on the streets of the city. I got as far as Washington Market, when the buyers were so plenty and eager to buy and so thoroughly blocked my way that I was compelled to stop and put a price on them, and though the price seemed enormous, they were soon disposed of at what I asked.[432]

He never went back to farming or the schoolroom.

Instead, Merritt hired hunters and bought birds killed by settlers throughout the Rock River valley and central Illinois. Railroads had reached the area in the mid-1850s, and he found that he could reliably collect, pack, and send birds to dealers in Chicago and New York. "In the summer of 1858, we shipped our first box of grouse to New York, ice packed," he explained. "The express was six dollars per hundred, and the box was packed so heavily with ice that, although the birds brought seventy-five cents per pair, we got but a few dollars out of it." He soon learned that if he gutted the birds, froze them solid, and packed them in reasonably air-tight containers, they would make the three-day trip safely without the expense of shipping so much ice:

> When we shipped in the early years, from 1861 to 1870, in the Spring, and cooled off the birds thoroughly on ice, we never lost but one shipment, and that was by transferring them from one town to another and letting the warm air work in. With light shoe boxes, as square as we could get them, we pressed them so closely, and they carried so much cold with them and retained it so well, that the birds could not damage in two or three days.[433]

Merritt's hunters shot anything that could be sold in Chicago or New York, often killing several hundred game birds or ducks in a day. Recalling the season of 1860, he wrote, "I sent the birds to Chicago and they sold by the dozen for $2.25 to $2.50, and as we bought them for a dollar they made us a good margin." Profits often varied for reasons he couldn't anticipate or control, and he had plenty of losing shipments. But on the whole, Merritt maintained a successful operation and a healthy income for half a century by sending Midwestern birds to eastern markets.[434]

Merritt was an astute businessman. For example, in the early 1860s he noticed that as migrations ended in the country and supplies fell, demand stayed the same in the city, so prices rose. Since he already had year-round icehouses to freeze birds, he wondered, "Could I keep those birds long enough on ice to secure the larger price? That I endeavored to do. I had practiced long enough keeping birds in that manner to know how long it would be safe to do so and have them marketable. I shipped everything up to April 20th, when snipe began to be fat, and packed away what came in afterward till close to the 10th of May." In May, wholesalers were happy to pay more to get birds that no one else had. He also profited from competing wholesalers: "Between the two New York markets, Washington and Fulton Street, there sprang up a great rivalry, and it would be a modest statement to say that between the two we got better returns than ever before."[435]

Those profits during the 1860s helped Merritt expand. Over the next two decades, his hunters spread through Illinois, across the Mississippi into Iowa, and up into Wisconsin to gather birds from local residents. He even built a small steamboat, the *Firefly*, on which his family and a few employees cruised the Upper Mississippi between La Crosse and the Quad Cities. They collected birds from local hunters up and down all its major tributaries and well into the hinterlands.

Midwestern birds killed by market hunters usually ended up in the hands of New York, Chicago, and Boston game dealers, whose lifestyle is now long forgotten. As the Civil War ended, Merritt started sending birds to

> the firm of Trimm & Sumner, on Washington Market, of whom Edward Sumner was the active manager of the game department. Beyond one or two mistakes, one of which at least was a heavy one, he was as good a man to trade with as I ever would want. I believe he was an honest trader, though his language was so vile and offensive you would at times feel justified to dispute it. He was a thoroughly competent judge of different kinds of game he handled. It was said of him that he would take a frozen snipe, chew it up without cooking, so he could tell whether it would thaw out sweet or not . . . Sumner talked everything, filthy beyond endurance or description, and never seemed to think it was out of place.[436]

But the undisputed ruler of the New York game market was Amos Robbins, of the firm of A. & E. Robbins. "In his department," Merritt recalled,

> his sway was absolute. All the managers of restaurants and hotels looked to him for their game. When he told them to drop one kind, it was done. When he told them to take up another it was done without asking why. He bought the best that money could buy and he would not put out on a good customer anything which he knew in any sense was inferior . . . He handled the market as a tradesman would handle his wares, selling them when and where they should be sold.[437]

Merritt recognized that his success depended on having these reliable partners. "With such an active dealer in each market and a commission man on the street, that person was fortunate who had the benefit of their united efforts. These three men all died within two or three years, not far from 1890, and bitterly have I felt their loss."[438]

Professional market hunters, of course, were not the only ones in the field with shotguns. The same forces that made commercial hunting possible enabled thousands of affluent city dwellers to hunt simply for fun.

Railroads blanket the Upper Midwest, from the *Sportsman's Gazetteer* (1877). *THE SPORTSMAN'S GAZETTEER AND GENERAL GUIDE*

Henry Herbert, writing at the end of the 1840s in New York, pointed out that just a few years earlier, "Travelling was, comparatively speaking, expensive; it was often necessary, in visiting out-of-the-way places, where the best sport was to be had, to hire private conveyances." But by 1848, on the East Coast, "All this is now changed [by] the rail-roads by which the country is everywhere intersected," enabling hunters to move about quickly and easily.[439]

Railroads reached the Midwest in the 1850s, and by the 1870s it seemed that every small town in the region had its railroad depot and telegraph office. In 1877, the editor of *Forest and Stream* published *The Sportsman's Gazetteer and General Guide*, a directory that included a 208-page list of the best hunting spots in the nation accessible by rail. It provided the train stations, hotel prices, and descriptions of birds and other game that could be shot at more than five hundred locations in the Ohio Valley, Midwest, and eastern Plains. Some of these sound surprising today, like the description of Flint, Michigan: "Bear, deer, wild turkey, quail, ruffed grouse and wild fowl shooting. Reached via the Flint and Pere Marquette Railroad. Gentlemen sportsmen will find accommodations at most of the farmhouses in the vicinity."[440]

Bird populations began to tumble under pressure from market gunners, sport hunters, and habitat loss from draining wetlands for farming. Conservation measures first appeared in the Midwest when hunting licenses were mandated late in the nineteenth century. The first states in the region to require licenses were Arkansas and Missouri (in 1875 and 1877), but these applied only to nonresidents. Most other states left hunting unregulated for another two decades, until they, too, began to require out-of-state residents to buy licenses. As late as 1904, half the states in the Midwest had no resident licensing requirement, since voters would not tolerate being told by the government what they could do on their own property.[441]

But even where they were mandated, licenses were largely ineffective since few hunters bothered to get them. In 1903, for example, just 218,008 hunting licenses were issued by all the eleven states in the Mississippi Flyway, when they had a combined population of nearly thirty-two million. Many local residents considered the birds to be their own property, just like their fields and woodlots. "Farmers would chase us with pitch forks and monkey wrenches and anything they could get hold of," said Dick Harker, an Iowa market gunner who recalled one encounter with a landowner. "He had a four tine pitch fork and was sure mad at me and told me to get out of that lake. I told him, 'You poor old fish, what's the matter with you? You don't own this lake. This is government water and I've got a perfect right here.' Then he was going to have me arrested and I told him to hop right to it."[442]

Professional hunters considered the fines for being caught without a license simply a cost of doing business. Harker recalled staying with a Minnesota farmer in 1895 or 1896 when the police raided the farmhouse and arrested him at gunpoint:

> They took us to the justice at Wilmer [Minnesota] and we plead guilty. He fixed up our fine. I think they fixed it at three hundred dollars, and I said, "How can you do that? The law says fifty dollars for the first offense." They went right back and came back with a fine of fifty dollars a piece. They took all our guns. We turned in our decoys and we got a good price on them. They gave us twenty dollars apiece for our guns. You could buy a new gun for eighteen dollars and mine was three or four years old then. In those days they [the police] took our goods and stuff in place of money. Amongst the three of us, we had to dig up thirty-five dollars.

He was still hunting without a license thirty-five years later.[443]

Behavior like this infuriated some educated, urban sport hunters, many of whom could see, as early as the 1870s, that the birds needed protection. Antipathy between market gunners and sportsmen intensified as the nineteenth century drew to a close and bird populations diminished. Harker described a clash near the Iowa-Minnesota border:

> One night when we got through hunting, we thought we'd hide our boats in the rushes. They found our boats that night and had an axe with them. We had a little strip boat that had been made by a boat builder. They took a couple of swipes and cut her wide open, then went to our other boat and took rocks that weighed three or four pounds and smashed it up for us. We didn't know who did this, but they were fellows who didn't like us. They were sportsmen from Minneapolis. They didn't want us there at all, because we could shoot and they couldn't hit anything. The boats were completely ruined. We built some more right there. They couldn't drive us out. They were ready to mob us one night. We didn't hear of this until some time after, but there was a gang all ready to mob us, but they never undertook it. One boy, some little fellow, took a shot at us.[444]

Violence between the two groups was common. Wealthy, educated hunters from the cities did everything in their power, legal and illegal, to acquire access to birds and preserve that access with conservation laws. Working-class market hunters and farmers in the country fought just as fiercely to protect their incomes and what they considered private property rights.[445]

The conflict had been growing for three decades when in 1884 the American Ornithologists' Union formed a committee, which consisted of many sportsmen and bird collectors, to gather "facts and statistics bearing upon the subject of the destruction of our birds." When the committee asked for grassroots data on migration and distribution, they were swamped with more than fourteen hundred responses. Two years later, the results of their findings were published as a special supplement to the journal *Science*.[446]

It exonerated sport hunters and ornithologists, suggesting that the former were in the forefront of bird protection and that the latter were too few in number to affect bird populations. All the specimens collected by professional ornithologists since American museums began, the AOU claimed, numbered 110,000 birds (modern scholars suggest closer to a million), and all those in private collectors' cabinets combined totaled no more than 300,000.

Instead, the AOU laid the blame for declining bird populations on five groups who could not have been more different from its own members. The first was, of course, "the 'professional' or 'market' gunners, by whom the ranks of the water-fowl are so fearfully thinned, and who often resort to any wholesale method of slaughter their ingenuity may be able to devise." But they also castigated African Americans, whose market stalls in the South contained thousands of songbirds, including flickers, robins, and even warblers destined for stew pots; immigrants, who executed "indiscriminate, often very quiet, slaughter" of birds with traps and poison grain; small boys, who they claimed robbed nests of eggs and killed birds indiscriminately with toy bows or pea-shooters; and "the dead-bird wearing gender."[447]

Although the affluent, educated AOU committee members denounced working-class immigrants, African Americans, and delinquent youth, it was women who received the most censure in their 1886 report. Examples were cited of eleven thousand songbirds collected for the millinery trade in a single trip to South Carolina; forty thousand terns collected one season on Cape Cod; and seventy thousand skins supplied to Long Island hatmakers in just four months. Nationwide, the AOU estimated that fashion trends were responsible for the deaths of five million birds annually. As we saw above, that was probably half the number killed by the nation's market hunters each year.[448]

The AOU report concluded that contemporary conservation laws were inadequate, unenforceable, and universally despised. "So apathetic is the public in all that relates to bird-protection," it said, "that prosecution under the bird-protection statutes requires, on the part of the prosecutor, a considerable amount of moral courage to face the frown of public opinion, the malignment of motive, and the enmities such prosecution is sure to engender." Despite its obvious biases on matters of class, race, ethnicity, and gender, the AOU's 1886 report sounded an alarm.[449]

Unfortunately, it went unheard. The 1,400 survey responses were left to gather dust, and effective conservation measures didn't begin for another decade when, in 1896, the first Audubon Society was formed in Massachusetts. It helped publicize horrific stories of bird slaughter for the millinery trade, prompting the establishment of fourteen other chapters. Then, in 1897, William Hornaday of the New York Zoological Society collected nearly two hundred questionnaires from observers around the country and concluded that, nationwide, nearly half of all American birds had perished since 1882.[450]

Devastating as Hornaday's statistical conclusion sounds, it lacks the emotional power of the brief comments made by local observers in response to his questionnaire:

> Detroit, Michigan: "Water fowl generally have decreased in number, almost to total extinction. Certainly not more than 1/4 as many seen now in Mich, as were seen 15 years ago."
>
> Milwaukee, Wisconsin: "Birds are decreasing fearfully in this locality. . . . Where, in the days of my boyhood, thirty-five years ago, orchards and woodlands were ringing with bird music, silence seems to reign supreme in these last years of the century . . . Of many of the birds named [our finest song and game birds] scarcely 1/4th are to be found now, while of numerous other species scarcely 1/10th or even 1/50th are to be met with to-day."

> Keokuk, Iowa: "Of birds of prey, one-half remain; of water fowl, one-third; game birds, one-tenth."
>
> Lawrence, Kansas: "Prairie chickens almost exterminated in eastern third of this state. Not one-thousandth as many as there were at one time. Wild turkey extinct in the state. Water birds—geese, ducks, Wilson's snipe, plover, etc.—about one-twentieth as many as formerly. Woodcock rare; ruffed grouse and wild pigeon extinct."[451]

Fifteen years later, in 1912, Hornaday performed another survey, which showed that twenty-five birds species had become extinct in at least one Midwestern state where they had once been common.[452] Three—the Carolina parakeet, passenger pigeon, and whooping crane—were entirely gone from the region. Although most birds killed for the millinery trade came from the Atlantic Coast or the South, a fair share came from the Midwest; the snowy egret, American white pelican, sandhill and whooping cranes, and trumpeter swan had all been decimated in the heart of the continent.[453]

If Midwestern market hunters averaged as many birds as Dick Harker did, they killed 2.8 million game birds and waterfowl annually between 1878 and 1918, or 112 million birds in all. But Hornaday's observers suggested that not market gunners but sportsmen and plume collectors were the chief cause of population declines. Assuming sportsmen contributed as much to the slaughter as market hunters did and that plume hunters killed half as much, the total would be more than 250 million birds killed in a single generation in the Mississippi Flyway.

Two hundred and fifty million birds? What were our ancestors thinking? People like Harker, Merritt, and Robbins, who shot, shipped, and sold the birds, as well as the millions who chewed and swallowed them or wore them on their hats, thought that their actions were unremarkable. Their beliefs, desires, and values prompted them to act in ways that we find nearly incomprehensible. They were not only caught in a web of economic and technological forces encouraging this great extermination, but their choices were validated by contemporary common sense.

Chief among the forces sanctioning their conduct was the ancient Judeo-Christian teaching that nature exists specifically for human exploitation. "No dogma taught by the present civilization," wrote John Muir in 1875, "seems to form so insuperable an obstacle in the way of a right understanding of the relations which culture sustains to wildness, as that which declares that the world was made especially for the uses of men." The book of Genesis taught Harker, Merritt, and millions of other late-nineteenth-century Americans that God had given them the natural world in order to "subdue it: and have dominion over the fish of the sea, and over the fowl of the air, and over every living thing that moveth upon the earth." I have heard Midwestern hunters justify their sport in exactly these terms in the twenty-first century.[454]

Only a few people disagreed. Pennsylvania German farmers looked instead to Genesis 2:15, which says that God "took the man, and put him into the Garden of Eden to till it and to care for it," and to Leviticus 25:23, where God declares, "The land shall not be sold forever. For the land is mine." They developed an environmental ethic based not on exploitation but on stewardship, which can still be seen at work today in Amish and Mennonite communities around the globe.[455]

The vast majority of early-twentieth-century Americans silently embraced the Old Testament version of man's place in nature and inserted themselves into it through the universal hero motif. They saw themselves as protagonists in a battle between wild nature and civilizing humanity, a conflict in which victory was not just sanctioned but demanded by God. Their beliefs, desires, and values didn't connect them intimately with the natural world but encouraged a hostile confrontation with it.

Their environmental ethic generally appalled their American Indian neighbors, who did not find nature threatening, nor did they view its inhabitants as enemies. "Not until the hairy man from the east came," wrote Lakota chief Luther Standing Bear, "and with brutal frenzy heaped injustices upon us and the families we loved, was it 'wild' for us. When the very animals of the forest began fleeing from his approach, then it was that for us the 'Wild West' began." Or as a modern Lakota woman, Jenny Leading Cloud, puts it, "White people see man as nature's master and conqueror, but Indians, who are close to nature, know better."[456]

Considering these differences, it's remarkable that a diametrical shift in values took place between 1886, when the AOU's initial plea for bird conservation fell on deaf ears, and 1918, when market gunning and plume hunting were finally outlawed. It was the result of an educational campaign based on the assumption that words were mightier than shotguns.

In February 1886, the same month that the AOU report appeared, the new editor of *Forest and Stream*, George Bird Grinnell, proposed the creation of the Audubon Society. Although the bird protection movement stumbled for another decade, in 1896 it took off again by focusing on the millinery trade, vastly expanding public education about birds, and advocating stricter hunting laws. As the new century dawned, public opinion began to swing behind these efforts, and by 1905 the AOU's model bird protection legislation was in force in twenty-eight states.[457]

This was during the Progressive Era, when the majority of voters expected their elected officials to aggressively address social problems. Bird protection took its place alongside women's suffrage, child labor, food safety laws, and other reforms. In 1900, the federal government passed the Lacey Act, ending interstate trade in illegally killed birds. Between 1901 and 1910, more than 1,300 game laws were passed at the state level, and by 1912, forty-one of the forty-eight states were paying salaried game wardens to enforce them.

When the federal government claimed the right to protect birds with the 1913 Migratory Bird Act, opponents stopped it in the courts, protesting that conservation was a local matter. So federal officials argued that migrations were international events and negotiated a treaty with Canada, the 1918 Migratory Bird Treaty, that has protected virtually all species of North American birds ever since.[458]

The great extermination ended only when millions of minds changed their concept of "normal." This shift in common sense occurred because between 1886 and 1918 leaders of environmental organizations orchestrated a massive marketing campaign in the media. John Burroughs became the nation's most popular nature writer and wrote the introduction for Neltje Blanchan's best-seller, *Bird Neighbors*. Educators taught an entire generation of schoolchildren the importance of conservation, and key decision makers were persuaded that it was in their best interest. By then, of course, many bird species had been reduced to population levels from which they would never recover, and the passenger pigeon and Carolina parakeet were gone forever.[459]

The Progressive Era conservation movement channeled Americans' thinking about birds into two main streams: science and sentiment. Trained biologists had marshaled data and recommended policy changes, much as they do today about water quality or climate change. At the same time, marketers and publishers kept the media supplied with romantic images and emotional stories about birds, and teachers exhorted millions of children to love their "feathered friends" (a phrase used in thirty-three book titles and more than a thousand newspaper articles between 1890 and 1920).[460]

As a result, today the vast majority of Americans interested in birds feel affection toward them, or even sometimes awe, in situations where their great-grandparents would have reached for the shotgun. Those who want to study birds seriously have 1,500 scientific journals and forty-five PhD programs to choose from. These days, the general public buys countless avian-themed T-shirts, calendars, coffee mugs, and other knickknacks, while academics mine more than sixty-five thousand technical articles in the Ornithological Worldwide Literature database. The two streams merge and overlap at major zoos, venues like the International Crane Foundation in Baraboo, Wisconsin, Web sites such as Cornell's AllAboutBirds.org, and smartphone apps such as Mitchell Waite's iBird.

11

Thinking about People

Although we go outdoors to watch birds, we mostly encounter ourselves. As novelist Anais Nin put it, "We see things not as they are but as we are."[461]

Hearing a distant squawk and looking up, my awareness is flooded by sights, sounds, and smells. My mind instantly transforms this stream of sensations into words, fits them into categories, and colors them with my memories, fears, hopes, and desires. I can't separate the wedge of geese overhead from those associations in my mind. I don't experience the geese as they are, but only through my own particular cultural and psychological filters.

Whether a person sees migrating geese as scientific specimens, spiritual brethren, quantifiable data, divine messengers, indications of approaching winter, or a potential meal (or even whether they mistake them for big ducks) depends on how their community speaks and what it values. The actions they take depend on those same influences. Whether I'm tempted to reach for binoculars or a shotgun, stop to pray, or merely smile in gratitude is determined by the shared beliefs, desires, and values of my culture. These change over time.

At my favorite birding spot on the edge of town, a dirt road winds through the woods to a scrubby point jutting into the Yahara River. As I reach the end of it and sweep the water with my binoculars, I stand next to a conical burial mound made by Hopewell people two thousand years ago. This little peninsula was special to them, too.

Across the bay, a two-hundred-foot panther effigy has been keeping silent watch for more than a thousand years. A mile downriver, where the Yahara empties into Lake Mendota, a flock of bird effigies overlooks the shoreline. Generations of hardwoods have grown old, tumbled to the ground, and melted back into the soft earth while these sacred bird mounds tried to balance heaven and hell.

Over the hill that rises behind me, a Yankee farmer attempted to drain the wetland one hundred years ago. His perpendicular ditches were no match for the gentle curves of the marsh as it meandered among alders, sumac, and cattails. Today, blue-winged teal and mallards nest where he dreamed of harvesting corn. After him, university students came here to study biology, and their professors lobbied to preserve the marsh. Since it was too wet, too buggy, and too far from downtown to have financial value, city officials agreed.

The same birds have returned here each spring for thousands of years. Great blue herons step gingerly through the shallows as they always have, and sandhill cranes still stop for the night on their way north. But what the humans here noticed and thought about birds, and what they did to them, has continually changed.

So the marsh that was once a holy place for shamanic ceremonies became a hunting ground, a family farm, a biology lab, and a conservation park. At each point in time, people's ideas prompted specific actions that reshaped this landscape. Humans have been reshaping the Midwest for at least twelve thousand years.

But in just the last two centuries, we have unintentionally brought the entire planet to the brink of disaster by altering its climate. To save ourselves from scorching droughts, increasingly violent storms, and the political and military calamities they'll produce, we need to change how we think. The Buddha taught that actions follow ideas "as surely as the wheel follows the footsteps of the ox." Changing the planet begins by changing our minds.

Each of us needs to examine which of our beliefs, desires, or values encourage us to act in ways that threaten the future, where those notions come from, who profits from them, and who suffers from them—to become mindful of the silent assumptions that we consider common sense. Only then can we imagine alternatives and make different choices.

Do we really need to burn millions of barrels of oil trucking food thousands of miles from field to table? Leave hundreds of millions of computers and televisions buzzing on standby twenty-four hours a day? Overheat and air-condition millions of homes and offices merely for personal comfort? Flush billions of gallons of perfectly clean water down our toilets? Drive our two-thousand-pound cars half a mile to the supermarket? Why do we think these actions are "normal"? Who wins and who loses when we perform them? How will our descendants judge them? What could we do differently?

Each of us needs to imagine how we might nurture the world as well as exploit it, how to tread as carefully in our own lives as a great blue heron does in the shallows. This starts by replacing our narcissism, fear, and greed with empathy, humility, and respect, and then by altering our daily habits to express those new values.

If enough of us can do this, perhaps our children's children will also feel inspired by the ancient Thunderbirds guarding the mouth of the Yahara, and sandhill cranes will still stalk the meadows beside them.

Sources and Acknowledgments

Ornithologist Elliott Coues called bibliography "a necessary nuisance and a horrible drudgery that no mere drudge could perform . . . It takes a sort of inspired idiot to be a good bibliographer, and his inspiration is as dangerous a gift as the appetite of the gambler or dipsomaniac—it grows with what it feeds upon, and finally possesses its victim like any other invincible vice." Coues knew from personal experience. He spent most of the 1870s compiling a five-hundred-page bibliography that listed every known publication about American birds.[462]

Nothing on that scale is attempted here. The following list contains just the primary sources and secondary studies quoted in the text or cited in the notes. Readers wishing to consult the originals will find most of the older works online at Google Books, the Internet Archive, the Biodiversity Heritage Library, or the Digital Public Library of America. Most of the recent academic articles are available through JSTOR or ProQuest.

The epigraph by the Buddha at the start of the book is adapted from the opening verses of the *Dhammapada*, in the Khuddakapatha chapter of the *Khuddaka Nikaya*. The Einstein quote is from Werner Heisenberg's *Encounters with Einstein* (Princeton University Press, 1983), page 114.

As this project gradually unfolded over twenty-five years, many people helped me in different ways. I'm delighted to thank the following friends and colleagues for particular assistance: Michael Gordon, Brad Gottschalk, Tod Highsmith, Kenneth Lange, and Louise Robbins, for reading and criticizing the entire manuscript; Bob Birmingham, for advice about Mississippian archaeology; John Broihahn, for taking me to Raddatz Rockshelter and explaining the work of his late colleagues Warren Wittry and Robert Hall; Bill Cronon, for a lunch conversation twenty years ago that helped me untangle the relationships among nature, consciousness, and culture, and for critiquing the chapter on bird names; Jane De Broux and Kathy Borkowski, for allowing me to repurpose material about Father Louis Nicolas and Henry Rowe Schoolcraft that first appeared in the *Wisconsin Magazine of History*; my son, Devin, for sharing his expertise in ecology and environmentalism; my daughter Rosalie, for helping with French translations and linguistic anthropology, and especially for asking if the dogs knew what time it was; my daughter Julia, for modeling a disciplined writing practice; Erika Janik, for always pushing me to think bigger; Jim Leary, for guidance on how to understand folklore about birds; archaeologist Janet Speth, for unpacking her collections and sharing her expertise in avian archaeology; my editor, Liz Wyckoff, who improved almost every paragraph of my prose; and last but not least, Mary Fiorenza, my best critic, biggest fan, and greatest support.

Sources

2011 National Survey of Fishing, Hunting, and Wildlife-Associated Recreation. National Overview. Arlington, VA: U.S. Fish and Wildlife Service, August 2012.

Alford, Thomas W. *Civilization.* Norman: Univ. of Oklahoma Press, 1936.

Allen, Elsa. "The History of American Ornithology Before Audubon." *Transactions of the American Philosophical Society,* n.s., 41, part 3 (1951): 387–591.

Allen, J. A. *Biographical Memoir, Elliott Coues 1842–1899, Read before the National Academy of Sciences, April 1909.* Washington: National Academy of Sciences, 1909.

Allouez, Claude. "Of the Mission St. Francois Xavier and the Nations Dependent Thereon." In *The Jesuit Relations and Allied Documents, 1610–1791; the Original French, Latin and Italian Texts, with English Translations and Notes . . . ,* vol. 55, edited by Reuben Gold Thwaites, 179–225. Cleveland: Burrows Bros. Co., 1896–1901.

———. "Recit d'un 3e Voyage Faict aux Ilinois." In *The Jesuit Relations and Allied Documents, 1610–1791; the Original French, Latin and Italian Texts, with English Translations and Notes . . . ,* vol. 60, edited by Reuben Gold Thwaites, 147–167. Cleveland: Burrows Bros. Co., 1896–1901.

American Ornithologists' Union.———. *AOU Check-list of North and Middle American Birds.* Online at http://checklist.aou.org/.

———. *Fifty Years' Progress of American Ornithology, 1883–1933,* rev. ed. Lancaster, PA: American Ornithologists' Union, 1933.

———. "Supplement: The Present Wholesale Destruction of Bird-Life in the United States." *Science* 7, no. 160 (February 26, 1886): 191–205.

"An Old Chippewa Chief." *Milwaukee Sentinel,* Aug. 16, 1895.

Audubon, John James. "Account of the Method of Drawing Birds Employed by J. J. Audubon, Esq., F.R.S.E. in a Letter to a Friend." *Edinburgh Journal of Science* 8, no. 1 (January 1828): 48–54.

———. "Audubon's Journey Up the Mississippi." *Illinois State Historical Society Journal* 35 (1942): 148–173.

———. *Ornithological Biography, or an Account of the Habits of the Birds of the United States of America . . .* Edinburgh: Adam Black, 1831–1839.

———. *Writings and Drawings*. New York: Library of America, 1999.

B., J.C. *Travels in New France*. Harrisburg, PA: Penn. Historical Commission, 1941.

Bacqueville de la Potherie. *Histoire de l'Amerique Septentrionale*. Paris, 1722.

Barrow, Mark V. *A Passion for Birds: American Ornithology After Audubon*. Princeton, NJ: Princeton University Press, 2000.

Barton, Benjamin Smith. *Fragments of the Natural History of Pennsylvania*. Philadelphia: Way & Groff, 1799.

Bartram, John. *Observation on the Inhabitants, Climate, Soil, Rivers, Productions, Animals, and Other Matters Worthy of Notice . . . in his Travels from Pensilvania to Onondago, Oswego and the Lake Ontario* . . . London, 1751.

Battalio, John T. *The Rhetoric of Science in the Evolution of American Ornithological Discourse*. Stamford, CT: Ablex, 1998.

Baughman, Ernest W. *Type and Motif-Index of the Folktales of England and North America*. Indiana Folklore Series no. 20. The Hague: Mouton & Co., 1966.

Becker, Marshall Joseph. "Birdstones: New Inferences Based on Examples from the Area around Waverly, New York." *Bulletin, Journal of the New York State Archaeological Association* 126 (2012): 1–31.

Bedini, Silvio. "The Scientific Instruments of the Lewis and Clark Expedition." *Great Plains Quarterly* 4, no. 1 (1984): 54–69.

Belting, Natalia Maree. "The Piasa: It Isn't a Bird!" *Journal of the Illinois State Historical Society* 66, no. 3 (Autumn 1973): 302–305.

Beltrami, Giacomo Constantin. *A Pilgrimage in America* . . . (London, 1828; edition quoted: Chicago: Quadrangle Books, 1962).

Benn, D. W. "Hawks, Serpents, and Bird-Men: Emergence of the Oneota Mode of Production." *Plains Anthropologist* 34 (1989): 233–260.

Bergen, Fanny D. "Animal and Plant Lore." *Memoirs of the American Folk-Lore Society* 7 (1899).

Berlin, Brent. *Ethnobiological Classification* (Princeton, N.J.: Princteon Univ. Press, 1992).

———. "The First Congress of Ethnozoological Nomenclature." *Journal of the Royal Anthropological Institute*, Vol. 12 (2006): 523-524.

———, and John P. O'Neill. "The Pervasiveness of Onomatopoeia in Aguaruna and Huambisa Bird Names." *Journal of Ethnobiology* 1/2 (December 1981): 238–261.

Berres, Thomas E. *Power and Gender in Oneota Culture*. Dekalb, IL: Northern Illinois University Press, 2001.

Biggs, William. *Narrative of . . . While He Was a Prisoner with the Kickapoo Indians*. New York: Garland Publishing, 1977.

Birmingham, Robert A., and Leslie E. Eisenberg. *Indian Mounds of Wisconsin*. Madison: University of Wisconsin Press, 2000.

Blagden, Sir Charles. "Letters . . . to Sir Joseph Banks on American Natural History and Politics, 1776–1780." *Bulletin of the New York Public Library* 7, no. 11 (1903): 407–446.

Blair, Emma Helen. *The Indian Tribes of the Upper Mississippi Valley and Region of the Great Lakes: As described by Nicolas Perrot* . . . Cleveland: Arthur H. Clark Company, 1911.

Blane, William N. *An Excursion through the United States and Canada during the Years 1822–23, by an Englishman*. London: Baldwin, 1824.

Boas, Franz. "The Mind of Primitive Man." *Annual Report of the Board of Regents of the Smithsonian Institution . . . for the Year Ending June 30, 1901*, 451–460. Washington, DC: GPO, 1902.

Bodle, Wayne. "Review of Gilbert Imlay: Citizen of the World by Wil Verhoven." *William and Mary Quarterly*, 3rd ser., 66, no. 2 (April 2009): 445–449.

Bonaparte, Charles Lucien. *American Ornithology, or the Natural History of Birds Inhabiting the United States, Not Given by Wilson*. Philadelphia: Carey & Lea, 1825.

———. *The Genera of North American Birds, and a Synopsis of the Species Found Within the Territory of the United States . . .* New York: J. Seymour, 1828.

———. "Observations on the Nomenclature of Wilson's Ornithology." *Journal of the Academy of Natural Sciences of Philadelphia* 4, no. 1 (1824): 163–200.

Boszhardt, Robert, and Geri Straub. *Hidden Thunder: Rock Art of the Upper Midwest*. Madison: Wisconsin Historical Society Press, 2016.

Botkin, B A., ed. *Lay My Burden Down: A Folk History of Slavery*. Chicago: University of Chicago Press, 1945.

Bourne, Edward G. *Narratives of the Career of Hernando de Soto in the Conquest of Florida as Told by a Knight of Elvas and in a Relation by Luys Hernandez de Beidma . . .* New York: Barnes and Co., 1904.

Brackenridge, Henry Marie. *Journal of a Voyage Up the River Missouri in 1811*. Ed. by Reuben G. Thwaites, in *Early Western Travels, 1748–1846*. Cleveland: Burrows Brothers, 1904.

Brasser, Ted J. "Self-Directed Pipe Effigies." *Man in the Northeast* 19. George's Mills, NH: Anthropological Research Center of Northern New England, 1980: 96–97.

Brewer, J. Mason. *Dog Ghosts and Other Texas Negro Folk Tales*. Austin: University of Texas Press, 1958.

Brooke, Michael, and T. R. Birkhead. *The Cambridge Encyclopedia of Ornithology*. New York: Cambridge University Press, 1991.

Brown, Cecil H. *Language and Living Things: Uniformities in Folk Classification and Naming*. New Brunswick, NJ: Rutgers University Press, 1984.

Brown, Charles E. *Ben Hooper Tales: Settler's Yarns from Green and Lafayette Counties, Wisconsin*. Madison: Wisconsin Folklore Society, 1944.

———. *Paul Bunyan Natural History*. Madison, Charles E. Brown, 1935.

———. *Sourdough Sam, Paul Bunyan's Illustrious Chief Cook and Other Famous Culinary Artists of His Great Pinery Logging Camps, Old Time Tales of Kitchen Wizards, the Big Cook Shanty, the Camp Fare, the Dinner Horn, and Sam's Cook Book*. Madison: Wisconsin Folklore Society, 1945.

Brown, James A. "The Archaeology of Ancient Religion in the Eastern Woodlands." *Annual Review of Anthropology* 26 (1997): 465–485.

———. "On the Identity of the Birdman of the Southeastern Ceremonial Complex." In *Ancient Objects and Sacred Realms: Interpretations of Mississippian Iconography*, edited by F. K. Reilly and J. Garber, 56–106. Austin: University of Texas Press, 2007.

———. "The Shamanic Element in the Hopewell Period Ritual." In *Recreating Hopewell*, edited by Douglas Charles and Jane Buikstra, 475–488. Gainesville: University Press of Florida, 2006.

———. "Spiro Art and Its Mortuary Contexts." In *Death and the Afterlife in Pre-Columbian America*, edited by Elizabeth Benson, 1–36. Washington, DC: Dumbarton Oaks, 1985):

Bruner, Jerome. *Acts of Meaning*. Cambridge: Harvard University Press, 1990.

Buffon, Georges-Louis Leclerc, Comte de. *Histoire Naturelle, Générale et Particulière, avec la Description du Cabinet du Roi*. Paris: L'Imprimerie Royale, 1749–1788, 36 volumes.

Bunnin, Nicholas, and Jiyuan Yu. *Blackwell Dictionary of Western Philosophy*. Malden, MA: Blackwell, 2004.

Burroughs, John. "The Invitation." In *The Writings of John Burroughs*, vol. 1. Boston: Houghton, Mifflin, 1905.

Butterfield, Consul W. *History of Brulé's Discoveries and Explorations, 1610–1626*. Cleveland, OH: Helman-Taylor, 1898.

Cadillac, Antoine de Lamothe. "Description de la Riviere du Detroit." In *Decouvertes et Etablissements des Francais dans l'Ouest et dans le Sud de l'Amerique Septentrionale (1614–1754): Memoires et Documents Originaux*, edited by Pierre Margry (Paris, 1871–1886), vol. 5, 192–194.

Campbell, Patrick. *Travels in the Interior Inhabited Parts of North America in the Years 1791 and 1792*. Toronto: Champlain Society, 1937.

Canfield, William H. *Outline Sketches of Sauk County*, vol. 1. Baraboo, WI: A.N. Kellogg, 1861–1874.

Carpenter, Roger. "Making War More Lethal: Iroquois vs. Huron in the Great Lakes Region, 1609 to 1650." *Michigan Historical Review* 27, no. 2 (Fall, 2001): 33–51 .

Carriere, Joseph Medard. *Tales from the French Folklore of Missouri*. Chicago: Northwestern University Press, 1937.

Cartier, Jacques. *The Voyages of . . .* Edited by Henry Percival Biggar. Ottawa: King's Printer, 1924.

Carver, Jonathan. *Journals of . . . and Related Documents, 1766–1770*. Minneapolis: Minnesota Historical Society Press, 1976.

———. *Travels through the Interior Parts of North-America in the Years 1766, 1767, and 1768*. London, the author, 1778.

———. *Travels Through the Interior Parts of North-America in the Years 1766, 1767, and 1768*. London: Walter and Crowder, 1778.

———. *Travels through the Interior Parts of North America, in the Years 1766, 1767, and 1768*. London: Printed for C. Dilly; H. Payne; and J. Phillips, 1781.

Casteneda, Pedro Reyes. *The Journey of Coronado: 1540–1542*. New York: A.S. Barnes, 1904.

Catesby, Mark. *The Natural History of Carolina, Florida, and the Bahama Islands . . .* London: 1731; edition quoted, Chapel Hill: University of North Carolina Press, 1985.

Catlin, George. *Letters and Notes on the Manners, Customs, and Condition of the North American Indians*. London: The author, 1841.

Charles, Douglas K., and Jane E. Buikstra, eds. *Recreating Hopewell*. Gainesvillle: University Press of Florida, 2006.

Cleland, Charles Edward. *The Prehistoric Animal Ecology and Ethnozoology of the Upper Great Lakes Region*. Ann Arbor: University of Michigan Museum of Anthropology, Anthropological Papers no. 29: 1966.

Clinton, De Witt. *Letters on the Natural History and Internal Resources of the State of New York, by Hibernicus*. New York: Bliss and White, 1822.

Coates, Benjamin. *A Biographical Sketch of the Late Thomas Say, Esq.: Read Before the Academy of Natural Sciences of Philadelphia, December 16, 1834*. Philadelphia: W. P. Gibbons, 1835.

Coffin, Tristram P. *Indian Tales of North America; an Anthology for the Adult Reader*. Philadelphia: American Folklore Society, 1961.

Cole, Henry Ellsworth. *A Standard History of Sauk County, Wisconsin*. Chicago: Lewis, 1918.

Collot, Georges-Henri-Victor. *A Journey in North America, Containing a Survey of the Countries Watered by the Mississippi, Ohio, Missouri, and Other Affluing Rivers . . .* Paris: A. Bertrand, 1826.

Cooper, William. "Description of a New Species of Grosbeak, Inhabiting the Northwestern Territory of the United States." *Annals of the Lyceum of Natural History of New York* 1, pt. 2 (1825): 219–222.

Coues, Elliott. "Bibliographical Appendix: List of Faunal Publications Relating to North American Ornithology." In *Birds of the Colorado Valley*. US Geological Survey of the Territories, Miscellaneous Publications no. 11, 567–1066. Washington, DC: Department of the Interior, Government Printing Office, 1878.

———. *Birds of the Colorado Valley*. Washington, DC: GPO, 1878.

———. "Dr. Coues' Column." *The Osprey, an Illustrated Monthly Magazine of Popular Ornithology* (November 1897), 39–40.

———. *Field Ornithology. Comprising a Manual of Instruction for Procuring, Preparing and Preserving Birds, and a Check List of North American Birds*. Salem, MA: Naturalists' Agency, 1874.

———. *Key to North American Birds, Containing a Concise Account of Every Species of Living and Fossil Bird at Present Known from the Continent North of the Mexican and United States Boundary*. Salem, MA: Naturalists' Agency, and New York: Dodd and Mead, 1872.

Crothers, George M. "Early Woodland Ritual Use of Caves in Eastern North America." *American Antiquity* 77, no. 3 (July 2012): 524–541.

Cuming, Fortescue. *Sketches of a Tour to the Western Country, through the States of Ohio and Kentucky* . . . in Reuben Gold Thwaites, *Early Western Travels*, vol. 4. Cleveland, OH: Arthur H. Clark, 1904.

Dablon, Claude. *Relation of What Occurred Most Remarkable in the Missions of the Fathers of the Society of Jesus, , . . . in the Years 1671 and 1672*. In *The Jesuit Relations and Allied Documents, 1610–1791; the Original French, Latin and Italian Texts, with English*

Translations and Notes . . . , vol. 56 (entire), edited by Reuben Gold Thwaites. Cleveland, OH: Burrows Bros. Co., 1896–1901.

Dana, Edmund. *Geographical Sketches on the Western Country, Designed for Emigrants and Settlers . . .* Cincinnati: Looker, Reynolds & Company Printers, 1819.

Dana, Richard Henry. *Two Years Before the Mast.* New York: Harper Brothers, 1842.

Davis, Mark H. "Market Hunters vs. Sportsmen on the Prairie." *Minnesota History* (Summer 2006): 48–60.

Debus, Allen G. *Man and Nature in the Renaissance.* Cambridge, UK: Cambridge University Press, 1978.

Denys, Nicolas. *Description and Natural History of the Coasts of North America*. Paris, 1672; edition quoted, Toronto: Champlain Society, 1908.

Deslandres, Dominique. "Exemplo Aeque ut Verbo: The French Jesuits Missionary World." In *The Jesuits: Cultures, Sciences, and the Arts, 1540-1773*, edited by John W. O'Malley, 258–273. Toronto: University of Toronto Press, 1999.

Diaz-Granados, Carol. "Marking Stone, Land, Body, and Spirit: Rock Art in Mississippian Iconography." In *Hero, Hawk, and Open Hand: American Indian Art of the Ancient Midwest and South*, edited by Richard F. Townsend, 139–149. Chicago: Art Institute of Chicago; New Haven: Yale University Press, 2004.

———, and James Richard Duncan. *The Petroglyphs and Pictographs of Missouri*. Tuscaloosa: University of Alabama Press, 2000.

———, and James R. Duncan, ed. *The Rock-Art of Eastern North America.* Tuscaloosa: University of Alabama Press, 2004.

Dickson, D. Bruce. "The Atlatl Assessed: A Review of Recent Anthropological Approaches to Prehistoric North American Weaponry." *Bulletin of the Texas Anthropological Society* 56 (1985): 1–38.

Donck, Adriaen van der. *A Description of the New Netherlands*. First published 1640; edition quoted, Syracuse, NY: Syracuse University Press, 1968.

Dorsey, Kurkpatrick. *The Dawn of Conservation Diplomacy: U.S.-Canadian Wildlife Protection Treaties in the Progressive Era.* Seattle: University of Washington Press, 1998.

Dorson, Richard M. *American Negro Folktales*. New York: Fawcett, 1967.

———. *Bloodstoppers & Bearwalkers: Folk Traditions of the Upper Peninsula*. Cambridge, MA: Harvard University Press, 1952.

———. *Jonathan Draws the Long Bow.* Cambridge, MA: Harvard University Press, 1946.

———. *Negro Folktales in Michigan*. Cambridge, MA: Harvard University Press, 1956.

Doty, James Duane. "Northern Wisconsin in 1820: A Report Made to Lewis Cass, Secretary of War, dated: Detroit, September 27, 1820." *Wisconsin Historical Collections* 7 (1876): 195–206.

Drooker, Penelope B. *The View from Madisonville: Protohistoric Western Fort Ancient Interaction Patterns*. Ann Arbor: Memoirs of the Museum of Anthropology, University of Michigan, no. 31, 1997.

Dunwoody, H.H.C. "Weather Proverbs." *Signal Service Notes* no. 9. Washington, DC: War Deptartment, 1883.

Dupre, Celine. "Cavelier de La Salle, Rene-Robert." *Dictionary of Canadian Biography.* Online at http://www.biographi.ca/.

Durand, Elias. *Memoir of the Late Thomas Nuttall.* Philadelphia: Sherman and Son, 1860.

Dutcher, William. "Report of the National Association of Audubon Societies . . . [including] a History of the Audubon Movement." *Bird-Lore* 7, no. 1 (January–February 1905): 45–57.

Eastman, Charles A. *Indian Boyhood.* New York: Doubleday, Page & Co., 1915.

Edmonds, Michael. *Out of the Northwoods: The Many Lives of Paul Bunyan.* Madison: Wisconsin Historical Society, 2009.

Emerson, Thomas E. "Materializing Cahokia Shamans." *Southeastern Archaeology* 22, no. 2 (Fall 2003): 138–140.

Emerson, Thomas E., Kenneth B. Farnsworth, Sarah U. Wisseman, and Randall E. Hughes. "The Allure of the Exotic: Reexamining the Use of Local and Distant Pipestone Quarries in Ohio Hopewell Pipe Caches." *American Antiquity* 78, no. 1 (January 2013): 48–67.

Erdoes, Richard, and Alfonso Ortiz, eds. *American Indian Myths and Legends.* New York: Knopf Doubleday, 2013.

Fabri, Nicolas-Claude de. "Notes on Specimens and Pictures of Specimens Brought by the Sieur de Monts from Acadia." Reprinted in *New American World: A Documentary History of North America to 1612,* vol. 4, edited by David Quinn, 357–358. New York: Arno Press and Hector Bye, 1979.

Faulkner, Charles H., Bill Deane, and Howard H. Earnest Jr. "A Mississippian Period Ritual Cave in Tennessee." *American Antiquity* 49, no. 2 (April 1984): 350–361.

Fenton, William N., and Merle H. Deardorff. "The Last Passenger Pigeon Hunts of the Cornplanter Senecas." *Journal of the Washington Academy of Sciences* 33, no. 10 (October 1943): 289–315.

Fire, John. *Lame Deer, Seeker of Visions.* New York: Simon and Schuster, 1994.

Fishel, Richard L. "Medicine Birds and Mill Creek-Middle Mississippian Interaction: The Contents of Feature 8 at the Phipps Site (13CK21)." *American Antiquity* 62, no. 3 (July 1997): 538–553 .

Flaherty, Gloria. *Shamanism and the Eighteenth Century.* Princeton, NJ: Princeton University Press, 1992.

Flint, Timothy. *Recollections of the Last Ten Years, Passed in Occasional Residences and Journeyings in the Valley of the Mississippi.* Boston: Cummings, Hilliard, and Co., 1826.

Fowke, Edith. *Folktales of French Canada.* Toronto: NC Press Ltd., 1979.

Fradkin, Arlene. *Cherokee Folk Zoology: The Animal World of a Native American People, 1700–1838.* New York : Garland, 1990.

Franquelin, Jean Baptiste Louis. *Carte de la France Septentrionalle: Contenant la Decouverte du Pays des Illinois.* Manuscript in the archives of the Depot des Cartes et Plans de la Marine in Paris.

Furst, Peter T. *Flesh of the Gods: The Ritual Use of Hallucinogens.* New York: Praeger, 1972.

Gagnon, Francois-Marc. Introduction to *The Codex Canadensis and the Writings of Louis Nicolas*. Tulsa, Oklahoma: Gilcrease Museum, and Montreal: McGill-Queens University Press, 2011.

———. "Biographical Notes on Louis Nicolas, Presumed Author of the Codex Canadensis." Article at *Library and Archives Canada* (www.collectionscanada.gc.ca/codex/026014-1200-e.html), viewed February 14, 2015.

Garriott, Edward B. *Weather Folk-Lore and Local Weather Signs*. Bulletin 294, U.S. Department of Agriculture (Washington, DC: GPO, 1903).

Gerstäcker, Friedrich. *Wild Sports in the Far West*. London: G. Routledge, 1855.

Goddard, Peter. "Science and Scepticism in the Early Mission to New France." *Journal of the Canadian Historical Association / Revue de la Société historique du Canada* 6, no. 1 (1995): 43–58.

Goldsmith, A. Sean, ed. *Ancient Images, Ancient Thought: The Archaeology of Ideology.* Proceedings of the Twenty-third Annual Conference, University of Calgary Archaeological Association, 1992.

Graustein, Jeannette E. *Thomas Nuttall, Naturalist: Explorations in America, 1808–1841*. Cambridge, MA: Harvard University Press, 1967.

Greber, N'omi B. "Enclosures and Communities in Ohio Hopewell." In *Recreating Hopewell*, edited by Douglas Charles and Jane Buikstra, 74–105. Gainesville: University Press of Florida, 2006.

Greenough, William P. *Canadian Folk-life and Folk-lore.* New York: Richmond, 1897.

Griffin, James B. *The Fort Ancient Aspect.* Ann Arbor: University of Michigan Press, 1943.

Gyles, John. *Memoirs of Odd Adventures, Strange Deliverances, etc.* . . . Boston: Kneeland and Greene, 1736.

Hahn, Roger. *The Anatomy of a Scientific Institution: the Paris Academy of Sciences, 1666-1803*. Berkeley: University of California Press, 1971.

Hall, Francis. *Travels in Canada and the United States in 1816 and 1817*. London: Longman, Hurst, Rees, Orme, & Brown, 1818.

Hall, Robert L. "An Anthropocentric Perspective for Eastern United States Prehistory." *American Antiquity* 42, no. 4 (October 1977): 499–518.

———. *An Archaeology of the Soul: North American Indian Belief and Ritual*. Champaign: University of Illinois Press, 1997.

Hallock, Charles. *The Sportsman's Gazetteer and General Guide* . . . New York: Forest and Stream,1877.

Hamon, J. Hill. "Bird Remains from a Sioux Indian Midden." *Plains Anthropologist* 6, no. 13 (1961): 208–212.

Hansel, A.K. "End moraines—The end of the glacial ride." Illinois State Geological Survey, Geobit 2. Champaign, IL: 2004.

Hanzeli, Victor Egon. *Missionary Linguistics in New France*. The Hague: Mouton, 1969.

Hariot, Thomas. *A Briefe and True Report of the New Found Land of Virginia* . . . London: 1588. Reprinted in *New American World: A Documentary History of North America to 1612*, vol. 3, edited by David Quinn, 139–155. New York: Arno Press and Hector Bye, 1979.

Harner, Michael. "Magic Darts, Bewitching Shamans, and Curing Shamans." In *Shamans Through Time*, edited by Jeremy Narby and Francis Huxley, 195–199. New York: Tarcher/Penguin, 2004.

Hassler, Donald M. "Henry Rowe Schoolcraft." In Patterson, Daniel, *Early American Nature Writers: A Biographical Encyclopedia*, 311–315. Westport, CT: Greenwood, 2008.

Haven, Janet. "Alexander Wilson." On the University of Virginia's Web site "Alexander Wilson, American Ornithologist" at http://xroads.virginia.edu/~public/wilson/bio.html.

Hayes, Kevin. *The Road to Monticello: The Life and Mind of Thomas Jefferson*. New York: Oxford University Press, 2008.

Hayne, David M. "Lom D'Arce de Lahontan, Louis-Armand de, Baron de Lahontan." *Dictionary of Canadian Biography Online* at www.biographi.ca/.

Hearne, Samuel. *Journey from Prince of Wales's Fort in Hudson's Bay to the Northern Ocean, 1769, 1770, 1771, 1772*. First published London, 1795; Toronto: Macmillan, 1958.

Hechenberger, David. "The Jesuits: History and Impact: From Their Origins Prior to the Baroque Crisis to Their Role in the Illinois Country." *Journal of the Illinois State Historical Society*100, no. 2 (Summer 2007): 85–109.

Heckewelder, John. *History, Manners, and Customs of the Indian Nations Who Once Inhabited Pennsylvania and the Neighboring States*. First published 1818; edition quoted, Philadelphia: Historical Society of Pennsylvania, 1876.

———. *A Narrative of the Mission of the United Brethren* . . . Cleveland: Burrows Bros. Co., 1907.

Henderson, Gwynn A., ed. *Fort Ancient Cultural Dynamics in the Middle Ohio Valley*. Madison, WI: Prehistory Press, 1992.

Hennepin, Louis. *Father Louis Hennepin's Description of Louisiana, Newly Discovered to the Southwest of New France* . . . Minneapolis: Minnesota Historical Society, 1938.

———. *A New Discovery of a Vast Country in America*. First published in Paris, 1697; edition quoted, Chicago, A. C. McClurg & Co., 1903.

Henry, Alexander. *Travels & Adventures in Canada and the Indian Territories between 1760 and 1776*. Toronto: Champlain Society, 1901.

Herbert, Henry William. *Frank Forester's Field Sports of the United States, and British Provinces, of North America*. New York: Springer & Townsend, 1849.

Hernández, Francisco. *Plantas y Animales de la Nueva Espana, y sus virtudes por Francisco Hernandez, y de Latin en Romance por Fr. Francisco Ximenez*. Mexico, 1615; Latin editions published 1628, 1648.

Herrick, Francis H. *Audubon the Naturalist: A History of His Life and Time*. New York: Appleton, 1917.

Herskovits, Melville. *The Myth of the Negro Past*. New York: Harper Brothers, 1941.

Hilger, M. Inez. *Chippewa Child Life and Its Cultural Background*. Bulletin 146, Smithsonian Institution Bureau of American Ethnology. Washington, DC: GPO, 1951.

Hinman, Bob. *The Golden Age of Shotgunning*. New York: Winchester Press, 1971.

Histoire Naturelle des Indes: The Drake Manuscript in The Pierpont Morgan Library. New York: W.W. Norton, 1996.

History of Crawford and Richland Counties, Wisconsin. Springfield, IL: Union Publishing Company, 1884.

Hodder, Ian, ed. *Archaeology as Long-Term History.* (Cambridge, UK: Cambridge University Press, 2009.

Hodge, Frederick. *Handbook of American Indians North of Mexico.* Washington, DC: GPO, 1912.

Holt, Julie Z. "Animal Exploitation and the Havana Tradition." In *Recreating Hopewell,* edited by Douglas Charles and Jane Buikstra, 446–463. Gainesville: University Press of Florida, 2006.

Hornaday, William. "The Destruction of Our Birds and Mammals: A Report on the Results of an Inquiry." Second Annual Report of the New York Zoological Society. New York: The Society, March 15, 1898, 77–127.

———. *Our Vanishing Wild Life: Its Extermination and Preservation.* New York: C. Scribner's Sons, 1913.

Hosteder, John A. "Toward Responsible Growth and Stewardship of Lancaster County's Landscape." *Pennsylvania Mennonite Heritage* 12, no. 3 (July 1989): 2–10.

Howard, James H. "Potawatomi Mescalism and Its Relationship to the Diffusion of the Peyote Cult." *Plains Anthropologist* Vol. 7, No. 16 (June 1962), 125-135.

Howison, John. *Sketches of Upper Canada, Domestic, Local, and Characteristic . . .* Edinburgh: Oliver & Boyd, 1821.

Hunter, John D. *Manners and Customs of Several Indian Tribes Located West of the Mississippi . . .* Minneapolis: Ross and Haines, 1957.

Imlay, Gilbert. *A Topographical Description of the Western Territory of North America.* London: Debrett, 1792.

International Society of Ethnobiology. 2006. *International Society of Ethnobiology Code of Ethics.* Online at http://ethnobiology.net/code-of-ethics/.

Irmscher, Christoph. "Audubon the Writer." Talk delivered at the July 25, 2007, premiere of the PBS American Masters documentary *John James Audubon: Drawn from Nature.* Online at www.audubonroyaloctavos.com/SITE/pages/Irmscher.html.

Irving, Laurence. "Stability in Eskimo Naming of Birds on Cumberland Sound, Baffin Island." *University of Alaska Anthropological Papers* 10 (1961/63): 1–12.

James, Edwin. *Account of an Expedition from Pittsburgh to the Rocky Mountains, under the Command of Major Stephen H. Long . . .* Philadelphia: H. C. Carey and I. Lea, 1823.

Jetté, René. *Dictionnaire généalogique des familles du Québec.* Montréal: Presses de l'Université de Montréal, 1983.

Johnsgard, Paul A. *Lewis and Clark on the Great Plains: A Natural History.* Lincoln: University of Nebraska Press, 2003.

Johnson, Oliver. "A Home in the Woods; Oliver Johnson's Reminiscences of Early Marion County." *Indiana Historical Society Publications* 16, no. 2 (1951): 143–234.

Joliet, Louis. "Letter 10 October 1674." Photostat of original manuscript, in the Wisconsin Historical Society Archives (SC 283).

Jones, Amanda. "Faunal Analysis: Reconstructing Subsistency and Seasonality at the Tremaine Site (47LC95)." *University of Wisconsin–LaCrosse Journal of Undergraduate Research* 17 (2014): 1–14.

Jones, Arthur Edward. "Catalogue of Jesuit Missionaries to New France and Louisiana, 1611 to 1800." In *The Jesuit Relations and Allied Documents, 1610–1791; the Original French, Latin and Italian Texts, with English Translations and Notes . . .*, vol. 71, 120–182, edited by Reuben Gold Thwaites. Cleveland, OH: Burrows Bros. Co., 1896–1901.

Jones, Arthur. *Rare or Unpublished Documents II: The Aulneau Collection, 1734–1745*. Montreal, 1893.

Jones, Rev. Peter (Kahkewaquonaby). *History of the Ojebway Indians*. London: A. W. Bennett, 1861.

Josselyn, John. *New-England's Rarities Discovered in Birds, Beasts, Fishes, Serpents, and Plants of That Country . . .* London: G. Widdowes, 1672; facsimile, Boston: William Veazie, 1865.

"Journal of the First Voyage of Columbus." In *The Northmen, Columbus and Cabot, 985–1503*, edited by Julius E. Olson and Edward Gaylord Bourne, 87–258. New York: Charles Scribner's Sons, 1906.

Joutel, Henri. *Journal Historique du Dernier Voyage que Feu M. de LaSale Fit dans le Golfe de Mexique . . .* Paris: Chez E. Robinot, 1713.

———. *A Journal of La Salle's Last Voyage*. New York: Corinth Books, 1962.

Joyce, Jeremiah. *A Familiar Introduction to the Arts and Sciences for the Use of Schools and Young Persons . . .* London: Longman Hurst, 1819.

Kalm, Pehr. "A Letter from . . . Containing a Particular Account of the Great Fall of Niagara." In Bartram, John, *Observation on the Inhabitants, Climate, Soil, Rivers, Productions, Animals, and Other Matters Worthy of Notice . . . in his Travels from Pensilvania to Onondago, Oswego and the Lake Ontario. . . .*, 79–94. London, 1751.

———. *Travels into North America Containing Its Natural History, and a Circumstantial Account of Its Plantations and Agriculture in General . . .* London, 1772.

Kalmbach, E. R. "In Memoriam: W. L. McAtee." *The Auk* 80 (1963): 474–485.

Keating, William H. *Narrative of an Expedition to the Source of St. Peter's River, Lake Winnepeek, Lake of the Woods, etc., Performed in the Year 1823 . . .* Philadelphia: Carey, 1824.

Keewaydinoquay. "The Legend of Miskwedo." *Journal of Psychedelic Drugs* vol. 11, no. 1/2 (Jan.–June 1979): 29.

Kehoe, Alice. "Robert Leonard Hall." *Encyclopedia of Global Archaeology*, edited by Claire Smith, 3195–3197. New York: Springer Verlag, 2014.

Kehoe, Alice Beck. *America before the European Invasions*. New York: Longman, 2002.

Kelly, Lucretia S. "Patterns of Faunal Exploitation at Cahokia." In *Cahokia: Domination and Ideology in the Mississippian World*, edited by Timothy R. Pauketat and Thomas Emerson, 69–88. Lincoln: University of Nebraska Press, 2000.

Kennedy, John F. "Remarks at a Dinner Honoring Nobel Prize Winners of the Western Hemisphere," online at Gerhard Peters and John T. Woolley, *The American Presidency Project*. www.presidency.ucsb.edu/ws/?pid=8623.

Kenney, James. "Journal of James Kenny, 1761–1763." *Pennsylvania Magazine of History and Biography* 37 (1913): 395–449.

Kip, William I. *The Early Jesuit Missions in North America*. New York, 1846; edition quoted, Carlisle, Mass.: Applewood Books, 2010.

Knox, John, Capt. *An Historical Journal of the Campaigns in North America for the Years 1758 1760*. Toronto: 1914.

Krech, Shepard. *Spirits of the Air: Birds and American Indians in the South*. Athens: University of Georgia Press, 2009.

La Barre, Weston. "Shamanic Origins of Religion and Medicine." *Journal of Psychedelic Drugs* vol. 11, no. 1/2 (Jan.–June 1979): 7–11.

———. "The Pre-Peyote Mescal Bean Cult." *The Peyote Cult*. 5th ed. Norman: University of Oklahoma Press, 1989, 105–109.

———. "Anthropological Perspectives on Hallucination and Hallucinogens," in *Hallucinations: Behavior, Experience, and Theory* edited by R.K. Siegel and L.J. West, 9–52. New York: John Wiley and Sons, 1975.

La Salle, Robert Cavelier de. *Relation of the Discoveries and Voyages of Cavelier de La Salle from 1679 to 1681: The Official Narrative*. Chicago: The Caxton Club, 1901.

Lafitau, Joseph-Francois. *Customs of the American Indians Compared with the Customs of Primitive Times*. Toronto: Champlain Society, 1974–1977.

Lahontan, Louis-Armand de Lom d'Arce, Baron de. "Memoirs of North America." In *New Voyages to North America*, vol. 1, edited by Reuben Gold Thwaites, 299–407. First published London, 1703; Chicago, McClurg & Co., 1905.

Lahontan, Louis-Armand, Baron de. "A Short Dictionary of the Most Universal Language of the Savages." *New Voyages to North America*, vol. 2, edited by Reuben Gold Thwaites, 732–750. First published London, 1703; Chicago: McClurg, 1905.

Lapham, Increase A. *The Antiquities of Wisconsin*. Washington: Smithsonian Institution, 1855.

Lavender, David. *The Way to the Western* Sea. Lincoln: University of Nebraska Press, 2001.

Le Clercq, Chretien. *New Relation of Gaspesia with the Customs and Religion of the Gaspesian Indians*. First published 1691; edition quoted, Toronto: Champlain Society, 1910.

Le Conte, John L., ed. *The Complete Writings of Thomas Say on the Entomology of North America*. New York, Balliere Brothers, 1859.

Le Page du Pratz, Antoine Simon. *The History of Louisiana, or of the Western Parts of Virginia and Carolina ...* London: 1774; edition quoted, New Orleans, 1947.

Le Page du Pratz, Antoine. *The History of Louisiana: Or of the Western Parts of Virginia and Carolina ...* London: Becket, 1774.

Leary, James P., ed. *Wisconsin Folklore*. Madison: University of Wisconsin Press, 1998.

Lepper, Bradley T. "The Newark Earthworks: Monumental Geometry and Astronomy at a Hopewellian Pilgrimage Center." In *Hero, Hawk, and Open Hand: American Indian Art of the Ancient Midwest and South*, edited by Richard F. Townsend, 73–82. Chicago: Art Institute of Chicago; New Haven: Yale University Press, 2004.

Levine, Lawrence. *Black Culture and Black Consciousness: Afro-American Folk Thought from Slavery to Freedom*. New York: Oxford University Press, 1977.

Levi-Strauss, Claude. *Totemism*. Boston: Beacon Press, 1963.

Lewis, Andrew J. *A Democracy of Facts: Natural History in the Early Republic*. Philadelphia: University of Pennsylvania Press, 2011.

Lewis, James Otto. *The Aboriginal Port-folio: A Collection of Portraits of the Most Celebrated Chiefs of the North American Indians*. Philadelphia: J. O. Lewis, 1835–1836.

Liette, Pierre-Charles de. "Memoir of DeGannes Concerning the Illinois Country." *Collections of the Illinois State Historical Library* 23 (1934): 302–395.

Long, John. *Voyages and Travels in the Years 1768–1788*. Chicago: Lakeside Press, 1922.

Loskiel, George Henry. *History of the Mission of the United Brethren Among the Indians in North America*. London: Printed for The Brethren's Society for the Furtherance of the Gospel, 1794.

Lovejoy, Arthur O. *The Great Chain of Being: A Study of the History of an Idea*. Cambridge, MA: Harvard University Press, 1936.

Loyola University Chicago. "Jesuits and the Sciences, 1540–1999." Online exhibit at www.lib.luc.edu/specialcollections/exhibits/show/jesuitsandthesciences.

Lyons, Henry G. *The Royal Society, 1660–1940*. Cambridge: The University Press, 1944.

Mails, E. Thomas. *Mystic Warriors of the Plains: The Culture, Arts, Crafts, and Religion of the Plains Indians*. First published 1972; edition quoted, New York: Mallard Press, 1991.

Marquette, Jacques. "Of the First Voyage Made by Father Marquette toward New Mexico, and How the Idea Thereof was Conceived." In *The Jesuit Relations and Allied Documents, 1610–1791; the Original French, Latin and Italian Texts, with English Translations and Notes . . .*, vol. 59, 86–163, edited by Reuben Gold Thwaites. Cleveland, OH: Burrows Bros. Co., 1896–1901.

———. "Of the Mission of Saint Esprit at Chagaouamigong Point." In *The Jesuit Relations and Allied Documents, 1610–1791; the Original French, Latin and Italian Texts, with English Translations and Notes . . .*, vol. 54, 148–194, edited by Reuben Gold Thwaites. Cleveland, OH: Burrows Bros. Co., 1896–1901.

Mather, Cotton. "Unpublished Manuscripts by Cotton Mather on the Passenger Pigeon." *The Auk* 55 (1938): 471–477.

Maximilian zu Wied-Neuwied. *Maximilian Prince of Wied's Travels in the Interior of North America, during the Years 1832–1834*. Illustrated by Karl Bodmer. London: Ackermann, 1843–1844.

Mayfield, Harold F. "Bird Bones Identified from Indian Sites at the West End of Lake Erie." *Condor* 74, no. 3 (1972): 344–347.

McAtee, W. L. *Nomina Abitera*. Privately printed, 1945.

———. *Nomina Abitera, Supplement*. Privately printed, 1954.

McAtee, Waldo Lee. *Local Names of Migratory Game Birds*. U.S. Dept. of Agriculture Miscellaneous Circular no. 13. Washington, DC: Government Printing Office, 1923.

McDermott, John Francis. *A Glossary of Mississippi Valley French, 1673–1850*. St. Louis, MO: Washington University Studies, n.s., no. 12, December 1941.

McKenna, Terence. "Time and Mind." Partial transcription of a taped workshop, May 26–27, 1990, New Mexico. Online at www.erowid.org/chemicals/dmt/dmt_writings3.shtml; viewed January 19, 2015.

McKenney, Thomas, and James Hall. *The History of the Indian Tribes of North America . . . Embellished with One Hundred and Twenty Portraits* . . . Philadelphia: E. C. Biddle, 1836–1844.

Merritt, H. Clay. *The Shadow of a Gun*. Chicago: Peterson Co., 1904.

Minnesota Historical Society. "Jeffers Petroglyphs." Web site at http://sites.mnhs.org/historic-sites/jeffers-petroglyphs/the-rock.

Mitchell, Roger. "Farm Talk from Marathon County." in *Wisconsin Folklore*, edited by James P. Leary, 89–105. Madison: University of Wisconsin Press, 1998.

Moring, John. *Early American Naturalists: Exploring the American West, 1804-1900*. New York: Cooper Square Press, 2002.

Morris, Thomas. *Miscellanies in Prose and Verse*. London: Printed for James Ridgway, 1791.

Mossman, Michael J. "H. R. Schoolcraft and Natural History on the Western Frontier, Part 4: Indian Agency Years with Thomas McKenney." *Passenger Pigeon* 55 (1993): 146–178.

Moulton, Gary, ed. *Journals of the Lewis and Clark Expedition*. Lincoln: University of Nebraska Press, 1993.

Muir, Edwin. *The Complete Poems of Edwin Muir*. Aberdeen, UK: Association for Scottish Literary Studies, 1991.

Mulford, Carla. *The Cambridge Companion to Benjamin Franklin*. New York: Cambridge University Press, 2009.

Musgrave, Jack. "Market Hunting in Northern Iowa." *Annals of Iowa* 26, no. 3 (January 1945): 173–197.

Narby, Jeremy, and Francis Huxley, ed. *Shamans through Time*. New York: Tarcher/Penguin, 2004.

Nicolas, Louis. *L'Algonquin au XVIIe Siècle: une Edition Critique, Analysée et Commentée de la Grammaire Algonquine du Père Louis Nicolas*, edited by Diane Daviault. Quebec: Presses de l'Universite du Quebec, 1994.

———. *The Codex Canadensis and the Writings of Louis Nicolas*, edited by Francois-Marc Gagnon. Tulsa, OK: Gilcrease Museum and Montreal: McGill-Queens University Press, 2011.

Nobles, Gregory H. "Ornithology and Enterprise: Making and Marketing John James Audubon's 'The Birds of America.'" *Proceedings of the American Antiquarian Society* 113, part 2 (2003): 267–302.

Nuttall, Thomas. *An Introduction to Systematic and Physiological Botany*. Cambridge, MA: Hilliard and Brown, 1830.

———. *Manual of the Ornithology of the United States and of Canada*. 2 volumes. Cambridge, MA: Hilliard and Brown, 1832 and 1834.

———. "Remarks and Inquiries Concerning the Birds of Massachusetts." *Memoirs of the American Academy of Arts and Sciences*, n.s. 1, 91–106. Cambridge: Charles Folsom, 1833.

———. "Travels into the Old Northwest: An Unpublished 1810 Diary." *Chronica Botanica* 14 (1950–51): 21–27, 42–77, 105–113.

O'Gorman, Jodie A. "Assessing Oneota Diet And Health: A Community and Lifeway Perspective." *Midcontinental Journal of Archaeology* 30, no. 1 (Spring 2005): 119–163.

Olson, Julius E., and Edward Gaylord Bourne, eds. *The Northmen, Columbus and Cabot, 985-1503*. New York: Charles Scribner's Sons, 1906.

O'Malley, John W. *The Jesuits: Cultures, Sciences, and the Arts, 1540-1773*. Toronto: University of Toronto Press, 1999.

Ord, George. "A Memoir of Thomas Say." In *The Complete Writings of Thomas Say on the Entomology of North America*, edited by John L. Le Conte, vii-xxi. New York: Balliere Brothers, 1859.

Overstreet, David. "Oneota Prehistory and History." *Wisconsin Archeologist* 78, nos. 1–2 (Jan.–Dec. 1997): 251–296.

Oviedo, Gonzalo Fernandez de. "Historia General y Natural de las Indias." Toledo, 1526. Excerpted in *New American World: A Documentary History of North America to 1612*, vols. 1–3, edited by David Quinn. New York: Arno Press and Hector Bye, 1979.

Owen, Mary Alicia. *Voodoo Tales, as Told among the Negroes of the South-west*. New York: G.P. Putnam's Sons, 1893.

Palmer, Theodore Sherman. *Chronology and Index of the More Important Events in American Game Protection. United States*. USDA Bureau of Biological Survey Bulletin no. 41. Washington, DC: U.S. Government Printing Office, 1912.

———. *Hunting Licenses: Their History, Objects, and Limitations*. USDA Biological Survey Bulletin no. 19. Washington, DC: US Department of Agriculture, 1904.

Parmalee, Paul W. "Additional Noteworthy Records of Birds from Archaeological Sites." *The Wilson Bulletin* 79, no. 2 (June 1967): 155–162.

———. "Animal Remains from the Raddatz Rockshelter, SK5, Wisconsin." *Wisconsin Archeologist* 40, no. 2 (1959): 83–90.

———. "The Avifauna from Prehistoric Arikara Sites in South Dakota." *Plains Anthropologist* 22 (August 1977): 189–222.

———. "Remains of Rare and Extinct Birds from Illinois Indian Sites." *The Auk* 75, no. 2 (April 1958): 169–176, 293–303.

Paterek, Josephine. *Encyclopedia of American Indian Costume*. Denver, CO: ABC-CLIO, 1994.

Patterson, Daniel, ed. *Early American Nature Writers: A Biographical Encyclopedia*. Westport, CT: Greenwood, 2008.

Pauketat, Timothy. *Cahokia: Ancient America's Great City on the Mississippi*. New York: Viking, 2009.

Pauketat, Timothy R. "The Forgotten History of the Mississippians." In *North American Archaeology*, edited by T. R. Pauketat and D. D. Loren, 187–212. Oxford, UK: Blackwell, 2005.

Pauketat, Timothy R., and Thomas Emerson. *Cahokia: Domination and Ideology in the Mississippian World*. Lincoln: University of Nebraska Press, 2000.

———. "The Ideology of Authority and the Power of the Pot." *American Anthropologist*, n.s., 93, no. 4 (December 1991): 919–941.

Pauketat, Timothy R., and D. D. Loren. *North American Archaeology.* Oxford, UK: Blackwell, 2005.

Penney, David W. "The Adena Engraved Tablets: A Study of Art Prehistory." *Midcontinental Journal of Archeology* 5, no. 1 (1980): 3–38.

———. "The Archaeology of Aesthetics." In *Hero, Hawk, and Open Hand: American Indian Art of the Ancient Midwest and South*, edited by Richard F. Townsend, 43–56. Chicago: Art Institute of Chicago; New Haven: Yale University Press, 2004.

Peters, Gerhard, and John T. Woolley. *The American Presidency Project.* At www.presidency.ucsb.edu/ws/?pid=8623.

Phillips, Philip. *Pre-Columbian Shell Engravings from the Craig Mound at Spiro, Oklahoma.* Cambridge, MA: Peabody Museum Press, 1975.

Pickens, A. L. "A Comparison of Cherokee and Pioneer Bird-Nomenclature." *Southern Folklore Quarterly* 7 (1943): 213–221.

Pike, Zebulon. *Journals of . . . with Letters and Related Documents*. Norman, OK: University of Oklahoma Press, 1966.

Pokagon, Chief [Simon]. *O-Gi-Maiv-Kiue Mit-I-Gwa-Ki (Queen of the Woods)* . . . Hartford, MI, 1899.

Pond, Peter. "1740–75: Journal of. . . ." *Wisconsin Historical Collections* 18 (1908): 314–354.

Porter, Charlotte M. *The Eagle's Nest: Natural History and American Ideas, 1812–1842.* Tuscaloosa, AL: University of Alabama Press, 1986.

Powell, Eric A. "Paleolithic Pastime." *Archaeology* 62, no. 1 (2009): unpaginated; Web edition at http://archive.archaeology.org/0901/etc/conversation.html.

Power, Susan C. *Feathered Serpents & Winged Beings: Early Art of the Southeastern Indians*. Athens: University of Georgia Press, 2004.

Prehistoric Animal Ecology and Ethnozoology of the Upper Great Lakes Region. University of Michigan Museum of Anthropology Anthropological Papers no. 29. Ann Arbor, 1966.

Pritchard, Evan T. *Bird Medicine: The Sacred Power of Bird Shamanism*. Rochester, VT: Bear & Co., 2013.

———. *Native American Stories of the Sacred*. Woodstock, VT: SkyLight Paths, 2005.

———. *No Word for Time: The Way of the Algonquin People*. San Francisco: Council Oak Books, 2001.

Proceedings of the 1989 Smoking Pipes Conference: Selected Papers. Rochester, N.Y.: Rochester Museum and Science Center Research Records no. 22,1992.

Puckett, Newbell Niles. *Folk Beliefs of the Southern Negro.* Chapel Hill: University of North Carolina Press, 1926.

Purchas, Samuel. *Hakluytys Posthumus, or Purchas His Pilgrimes*. London, 1625; edition quoted, Glasgow, 1906.

Quaife, Milo M. *Chicago and the Old Northwest, 1673–1835*. Chicago: University of Chicago Press, 1913.

———, ed. *The Western Country in the Seventeenth Century: The Memoirs of Lamothe Cadillac and Pierre Liette*. Chicago: Lakeside Press, 1947.

Quinn, David B., ed. *New American World: A Documentary History of North America to 1612*. New York: Arno Press and Hector Bye, 1979.

Radin, Paul. *Winnebago Hero Cycles: A Study in Aboriginal Literature*. Baltimore: Waverly Press, 1948.

———. *The Winnebago Tribe*. Washington, DC: 37th Annual Report of the Bureau of American Ethnology, 1923.

Radisson, Pierre Esprit. *Voyages of . . . Being an Account of His Travels and Experiences among the North American Indians, from 1652 to 1684*. New York: P. Smith, 1943.

Randolph, Vance. *We Always Lie to Strangers: Tall Tales from the Ozarks*. New York: Columbia University Press, 1951.

Rasles, Sebastien. "Letter from Father Sebastien Rasles, Missionary of the Society of Jesus in New France, to Monsieur his Brother." In *The Jesuit Relations and Allied Documents, 1610–1791; the Original French, Latin and Italian Texts, with English Translations and Notes . . .*, vol. 67, 131–228, edited by Reuben Gold Thwaites. Cleveland, OH: Burrows Bros. Co., 1896–1901.

———. "The Wanderings of Father Rasles, 1689–1723." In William Kip, *Early Jesuit Missions in North America*, 21–66. New York, 1846; edition quoted, Applewood Books, 2010.

Reilly, F. Kent. "People of Earth, People of Sky: Visualizing the Sacred in Native American Art of the Mississippian Period." In *Hero, Hawk, and Open Hand: American Indian Art of the Ancient Midwest and South*, edited by Richard F. Townsend, 125–137. Chicago: Art Institute of Chicago; New Haven: Yale University Press, 2004.

———. "The Petaloid Motif: A Celestial Symbolic Locative in the Shell Art of Spiro." In *Ancient Objects and Sacred Realms: Interpretations of Mississippian Iconography*, edited by F. Kent Reilly and James Garber, 39–55. Austin: University of Texas Press, 2007.

Reilly, F. K., and J. Garber, eds. *Ancient Objects and Sacred Realms: Interpretations of Mississippian Iconography*. Austin: University of Texas Press, 2007.

Relation of the Discoveries and Voyages of Cavelier de La Salle from 1679 to 1681, the Official Narrative. Edited and translated by Melville B. Anderson. Chicago: Caxton Club, 1901.

"Reply to the Preceding Communication; Containing Observations on Certain Species of Game Birds, and on the Names by Which They Are Distinguished among Sportsmen." *Medical Repository* 8, no. 2 (1804): 124–128.

Rhodes, Richard. *John James Audubon: The Making of an American*. New York: Knopf, 2004.

Rigal, Laura. "Empire of Birds: Alexander Wilson's American Ornithology." *Huntington Library Quarterly* 59, no. 2/3 (1996): 232–236.

Ripinsky-Naxon, Michael. *The Nature of Shamanism: Substance and Function of a Religious Metaphor*. Albany, N.Y.: SUNY Press, 1993.

Romain, William F. *Shamans of the Lost World: A Cognitive Approach to the Prehistoric Religion of the Ohio Hopewell*. New York: Rowman & Littlefield, 2009.

Rupp, William J. "Bird Names and Bird Lore Among the Pennsylvania Germans." *Proceedings and Addresses, Pennsylvania-German Society* 52, no. 2 (1947).

Sagard, Gabriel. *Histoire du Canada et Voyages que les Frères Mineurs Recollects y Ont Faicts pour la Conversion des Infidèles depuis l'an 1615*. Paris: Libr. Tross, 1866.

———. *The Long Journey to the Country of the Hurons by Father Gabriel Sagard*. Toronto: The Champlain Society, 1939.

Sagard, Gabriel Theodat. "Dictionnaire de la Langue Huronne." First published Paris, 1632; included at the end of vol. 4 of his *Histoire du Canada et Voyages* . . . Paris: Librarie Tross, 1866, unpaginated, 143 pages.

Salzer, Robert. "Wisconsin Rock Art." *Wisconsin Archeologist* 78, nos. 1–2 (1997): 48–77.

Salzer, Robert, and Grace Rajnovich. *The Gottschall Rockshelter: An Archaeological Mystery*. St. Paul, MN: Prairie Smoke Press, 2000.

Schaeffer, Claude E. "Bird Nomenclature and Principles of Avian Taxonomy of the Blackfeet Indians." *Journal of the Washington Academy of Sciences* 40 (1950): 37–46.

Schoolcraft, Henry. *Narrative Journals of Travels from Detroit Northwest through the Great Chain of American Lakes to the Sources of the Mississippi River in the Year 1820*. First published Albany, NY, 1821; edition quoted, New York: Arno Press, 1970.

Schultes, Richard Evans. "An Overview of Hallucinogens in the Western Hemisphere." In Peter T. Furst, *Flesh of the Gods: The Ritual Use of Hallucinogens*, 2–54. New York: Praeger, 1972.

Schultes, Richard Evans, and Albert Hofmann. *The Botany and Chemistry of Hallucinogens*, 2nd ed. Springfield, IL: Charles C Thomas, 1980.

Schultes, Richard Evans, Albert Hofmann, and Christian Ratsch. *Plants of the Gods: Their Sacred, Healing, and Hallucinogenic Powers*. Rochester, VT: Healing Arts Press, 2001.

Schweinsberger, Sanchia. "Bone Flutes and Whistles Found in Ohio Valley Sites." *Proceedings of the Indiana Academy of Sciences* 54 (1949): 28–33.

Seeman, Mark F. "Hopewell Art in Hopewell Places." In *Hero, Hawk, and Open Hand: American Indian Art of the Ancient Midwest and South*, edited by Richard F. Townsend, 57–72. Chicago: Art Institute of Chicago; New Haven: Yale University Press, 2004.

Serjeantson, Dale. *Birds*. Cambridge Manuals in Archaeology series. New York: Cambridge University Press, 2009.

Severo, Richard. "Roger Peterson, 87, the Nation's Guide to the Birds, Is Dead." *New York Times*, July 30, 1996.

Shea, John G. *Early Voyages Up and Down the Mississippi, by Cavelier, St. Cosme, Le Sueur, Gravier, and Guignas*. Albany, NY: Joel Munsell, 1861.

Siegel, R.K. and L.J. West. *Hallucinations: Behavior, Experience, and Theory*. New York: John Wiley and Sons, 1975.

Smith, Bruce D. "Middle Mississippi Exploitation of Animal Populations: A Predictive Model." *American Antiquity* 39, no. 2 (April 1974): 274–291.

Smith, Claire, ed. *Encyclopedia of Global Archaeology*. New York: Springer Verlag, 2014.

Smith, James. *An Account of the Remarkable Occurences in the Life and Travels of . . . during his Captivity with the Indians*. Cincinnati: Robert Clarke & Co., 1870.

Smith, Michael. *A Geographical View of the Province of Upper Canada* . . . (Hartford, Conn.:, John Russell, 1813.

Speck, Frank G., and John Witthoft. "Some Notable Life-Histories in Zoological Folklore." *Journal of American Folklore* 60, no. 238 (1947): 345–349.

Squier, E. G., and E. H. Davis. *Ancient Monuments of the Mississippi Valley*. Washington: Smithsonian Institute Press, 1998.

Standing Bear, Luther. *Land of the Spotted Eagle*. Boston: Houghton Mifflin, 1931.

———. *My Indian Boyhood*. Lincoln: University of Nebraska Press, 1931.

———. *My People, the Sioux*. Boston: Houghton Mifflin, 1928.

Stanton, William. *American Scientific Exploration, 1803-1860*. American Philosophical Society Web site at www.amphilsoc.org/guides/stanton/0335.htm.

Stearns, Raymond P. *Science in the British Colonies of America*. Urbana: University of Illinois Press, 1970.

Steckley, John L. *Words of the Huron*. Waterloo, ON: Wilfrid Laurier University Press, 2007.

Stevenson, Katherine Phyllis. *Oneota Subsistence-Related Behavior in the Driftless Area: A Study of the Valley View Site near La Crosse, Wisconsin*. PhD diss., University of Wisconsin, 1985.

Story of American Hunting and Firearms. New York: Outdoor Life-Dutton, 1976.

Strachey, William. *The Historie of Travell into Virginia Britania (1612)*. London: The Hakluyt Society, 1953.

Strassman, Rick. *DMT: The Spirit Molecule: A Doctor's Revolutionary Research into the Biology of Near-Death and Mystical Experiences*. Rochester, VT: Park Street Press, 2001.

Strothers, David M., and Timothy J. Abel. 1993. "Archaeological Reflections on the Late Archaic and Early Woodland Time Periods in the Western Lake Erie Region." *Archaeology of Eastern North America* 21 (Fall 1993): 25–109.

Stroud, Patricia T. *Emperor of Nature: Charles-Lucien Bonaparte and His World*. Philadelphia: University of Pennsylvania Press, 2000.

Swanton, John R. "Newly Discovered Powhatan Bird Names." *Journal of the Washington Academy of Sciences* 24 (1934): 96–99.

Szabo, Joyce M. *Prehistoric Bannerstones and Birdstones of the Midwest with a Catalogue of the Collection from the Ohio Historical Society*. Master's thesis, Vanderbilt University, May 1978.

Tabeau, Pierre-Antoine. *Narrative of Loisel's Expedition to the Upper Missouri*. Norman: University of Oklahoma Press, 1939.

Taché, Karine. "New Perspectives on Meadowood Trade Items." *American Antiquity* 76, no. 1 (January 2011): 41–79.

Tallant, Robert. *Voodoo in New Orleans*. New York: Macmillan, 1946.

Taylor, Colin. *Wapah' a: The Plains Feathered Head-Dress*. Wyk auf Foehr, Germany: Verlag für Amerikanistik, 1996.

Taylor, Rose S. "Peter Schuster: Dane County Farmer (Part II)." *Wisconsin Magazine of History* 28, no. 4 (June 1945): 431–454.

The Sportsman's Companion, or An Essay on Shooting . . . [with] Directions to Gentlemen for the Treatment and Breaking Their Own Pointers and Spaniels . . . (Burlington, NJ: Isaac Neale, 1791.

Theler, James L., and Robert F. Boszhardt. *Twelve Millennia: Archaeology of the Upper Mississippi River Valley*. Iowa City: University of Iowa Press, 2003.

Thevenot, Melchisedec. "Découverte de quelques pays et nations de l'Amerique Septentrionale [par le P. Marquette]." In *Recueil de Voyages de Mr Thevenot. . . .* Paris: Estienne Michallet, 1681.

Thomas, Daniel L., and Lucy B. Thomas. *Kentucky Superstitions*. Princeton, NJ: Princeton University Press, 1920.

Thomson, Keith. *Jefferson's Shadow: The Story of His Science.* New Haven, CT: Yale University Press, 2012.

Thoreau, Henry. *Journal*. Boston: Houghton Mifflin, 1906.

Thoreau, Henry David. "Natural History of Massachusetts." In *Excursions*, 37–72. Boston: Ticknor and Fields, 1863.

Thwaites, Reuben G., ed. *Early Western Travels, 1748-1846.* Cleveland: Arthur H. Clark, 1904.

———, ed. *The Jesuit Relations and Allied Documents, 1610-1791; the Original French, Latin and Italian Texts, with English Translations and Notes...* Cleveland: Burrows Bros. Co., 1896–1901.

Tidemann, Sonia, and Andrew Gosler. *Ethno-Ornithology: Birds, Indigenous Peoples, Culture and Society*. Washington, DC: Earthscan, 2010.

Tooker, Elizabeth. *Native North American Spirituality of the Eastern Woodlands*. New York: Paulist Press, 1979.

Townsend, Earl C. *Birdstones of the North American Indian: A Study of These Most Interesting Stone Forms, the Area of Their Distribution, Their Cultural Provenience, Possible Uses, and Antiquity.* Indianapolis: Privately printed, 1959.

Townsend, Richard F. "American Landscapes, Seen and Unseen." In *Hero, Hawk, and Open Hand: American Indian Art of the Ancient Midwest and South*, edited by Richard Townsend, 15–36. Chicago: Art Institute of Chicago; New Haven: Yale University Press, 2004).

———, ed. *Hero, Hawk, and Open Hand: American Indian Art of the Ancient Midwest and South*. Chicago: Art Institute of Chicago; New Haven: Yale University Press, 2004.

Tremblay, Guy. *Louis Nicolas, Sa Vie et Son Oeuvre; les Divers Modes de Transport des Indiens Americains*. Unpublished master's thesis, University of Montreal, July 1983.

Trumbull, Gurdon. *Names and Portraits of Birds Which Interest Gunners, with Descriptions in Language Understanded of the People.* New York: Harper Brothers, 1888.

Tyler, Hamilton. *Pueblo Birds and Myths*. Norman: University of Oklahoma Press, 1979.

U.S. Census Bureau. "Table 1. Urban and Rural Population: 1900 to 1990." Washington, DC: U.S. Census Bureau, October 1995. Viewed May 3, 2015, at www.census.gov/population/censusdata/urpop0090.txt.

US Census Office. *Report on Population of the United States at the Eleventh Census, 1890, Part II.* Washington, DC: GPO, 1895–1897.

U.S. Fish & Wildlife Service. *Birding in the United States: A Demographic and Economic Analysis. Addendum to the 2006 National Survey of Fishing, Hunting, and Wildlife-Associated Recreation, Report 2006-4*. Arlington, VA: US Fish and Wildlife Service, July 2009.

Ubelaker, Douglas H., and Waldo R. Wedel. "Bird Bones, Burials, and Bundles in Plains Archaeology." *American Antiquity* 40, no. 4 (October 1975): 444–452.

Van Der Donck, Adriaen. "Description of New-Netherlands." *Collections of the New York Historical Society*, 2nd ser., 1 (1841): 125–242.

Vivier, Louis. "Letter from Father Vivier, Missionary among the Ilinois, to Father * * *." In *The Jesuit Relations and Allied Documents, 1610–1791; the Original French, Latin and Italian Texts, with English Translations and Notes . . .*, vol. 69, 141–148, edited by Reuben Gold Thwaites. Cleveland, OH: Burrows Bros. Co., 1896–1901.

Von Gernet, Alexander. "Hallucinogens and the Origins of the Iroquoian Pipe/Tobacco/Smoking Complex," in *Proceedings of the 1989 Smoking Pipes Conference: Selected Papers*, 171–185. Rochester , NY: Rochester Museum and Science Center Research Records no. 22,1992.

———. "New Directions in the Construction of Prehistoric Amerindian Belief Systems." In *Ancient Images, Ancient Thought: The Archaeology of Ideology*, edited by Sean A. Goldsmith, 133–140. Proceedings of the Twenty-third Annual Conference, University of Calgary Archaeological Association, 1992.

———, and Peter Timmins. "Pipes and Parakeets: Constructing Meaning in an Early Iroquoian Context." In *Archaeology as Long-Term History*, edited by Ian Hodder, 31–42. Cambridge, UK: Cambridge University Press, 2009.

Wagner, Mark, Mary R. McCorvie, and Charles Swedlund. "Mississippian Cosmology and Rock Art at the Millstone Bluff Site in Southern Illinois." In *The Rock Art of Eastern North America*, edited by Carol Diaz-Granados and James Duncan, 42–64. Tuscaloosa: University of Alabama Press, 2004.

Walthall, John, and Thomas Emerson. "Indians and French in the Midcontinent." In *Calumet and Fleur-de-Lys: Archaeology of Indian and French Contact in the Midcontinent*, edited by John A. Walthall and Thomas E. Emerson, 1–13. Washington, DC: Smithsonian Institution Press, 1992.

Warkentin, Germaine. "Aristotle in New France: Louis Nicolas and the Making of the Codex Canadensis." *French Colonial History* 11 (2010): 71–107.

Warren, Leonard. *Maclure of New Harmony: Scientist, Progressive Educator, Radical Philanthropist*. Bloomington: Indiana University Press, 2009.

Watts, Alan. *Nature, Man and Woman*. New York: Knopf Doubleday, 2012; first published 1958.

Webb, William S., and Charles E Snow. *The Adena People*. Lexington: University of Kentucky Reports in Anthropology and Archaeology 6 (1945).

Webb, William S., and Raymond S. Baby. *The Adena People, no. 2*. Columbus: Ohio Historical Society, 1957.

Weld, Isaac Jr. *Travels through the States of North America, and the Provinces of Upper and Lower Canada, in the Years 1795, 1796 and 1797*. London: J. Stockdale, 1799.

White, Hayden. *The Content of the Form*. Baltimore: Johns Hopkins Press, 1987.

"Who Were the Hopewell?" *Archaeology: A Publication of the Archaeological Institute of America* (2009), at archive.archaeology.org/online/features/hopewell/who_were_hopewell.html

Whorf, Benjamin Lee. *Language, Thought and Reality: Selected Writings of Benjamin Lee Whorf.* Cambridge: MIT Press, 1956.

Wilbert, Johannes. "Tobacco and Shamanic Ecstasy among the Warao Indians of Venezuela." In Peter T. Furst, *Flesh of the Gods: The Ritual Use of Hallucinogens*, 55–83. New York: Praeger, 1972.

———. *Tobacco and Shamanism in South America*. New Haven, CT: Yale University Press, 1987.

Williams, Mike. *Prehistoric Belief: Shamans, Trance and the Afterlife.* Stroud, Gloucestershire, UK: The History Press, 2010.

Williams, Roger. *A Key into the Language of America*. London, 1643. Reprinted in vol. 1 of *Publications of the Narragansett Club* (1866).

Willis, David, Charles Scalet, and Lester D. Flake. *Introduction to Wildlife and Fisheries*. New York: W.H. Freeman, 2009.

Wilson, Alexander. *American Ornithology.* London: Chatto and Windus, 1876.

Winship, George P., ed. *The Journey of Coronado, 1540–1542, from the City of Mexico to the Grand Canon of the Colorado and the Buffalo Plains of Texas, Kansas and Nebraska, As Told by Himself and His Followers*. New York: Barnes & Co., 1904.

Wittry, Warren L. "A Raven Headdress from Sauk County, Wisconsin." *Wisconsin Archeologist* 43, no. 4 (1963): 87–94.

Worcester, Donald, and Thomas F. Schilz. "The Spread of Firearms among the Indians on the Anglo-French Frontiers." *American Indian Quarterly* 8, no. 2 (Spring 1984): 103–115.

Wrong, George McKinnon, and William Stewart Wallace, eds. *Review of Historical Publications Relating to Canada.* Toronto: University of Toronto Press, 1897.

Young, Martha Strudwick. *Plantation Bird Legends*. New York: Appleton, 1902; edition quoted, 1916.

Zeisberger, David. *History of the Northern American Indians*. Columbus: Ohio State Archaeological and Historical Society, 1910.

Ziser, Michael. "Alexander Wilson." In *Early American Nature Writers: A Biographical Encyclopedia,* edited by Daniel Patterson, 388–393. Westport, CT: Greenwood, 2008.

Notes

Chapter 1

1. John Burroughs, "The Invitation," *The Writings of John Burroughs*, vol. 1 (Boston: Houghton, Mifflin, 1905), 218.
2. U.S. Fish & Wildlife Service, *Birding in the United States: A Demographic and Economic Analysis. Addendum to the 2006 National Survey of Fishing, Hunting, and Wildlife-Associated Recreation*, report 2006-4 (Arlington, VA: U.S. Fish and Wildlife Service, July 2009); *2011 National Survey of Fishing, Hunting, and Wildlife-Associated Recreation. National Overview* (Arlington, VA: U.S. Fish and Wildlife Service, August 2012).
3. Genesis 1:28 (New International Version).
4. William Fenton and Merle Deardorff, "The Last Passenger Pigeon Hunts of the Cornplanter Senecas," *Journal of the Washington Academy of Sciences* 33 (1943): 293.
5. Benjamin Lee Whorf, *Language, Thought and Reality: Selected Writings ...* (Cambridge: MIT Press, 1956), 213, 240.
6. Ibid., 218.
7. International Society of Ethnobiology, *International Society of Ethnobiology Code of Ethics* (with 2008 additions), at http://ethnobiology.net/code-of-ethics (2006).

Chapter 2

8. Michael Brooke and T. R. Birkhead, *The Cambridge Encyclopedia of Ornithology* (New York: Cambridge University Press, 1991), 62; Dale Serjeantson, *Birds* (Cambridge Manuals in Archaeology; New York: Cambridge University Press, 2009), 335.
9. James L. Theler and Robert F. Boszhardt, *Twelve Millennia: Archaeology of the Upper Mississippi River Valley* (Iowa City: University of Iowa Press, 2003), 37.
10. Ibid., 11–12.
11. Paul W. Parmalee, "Animal Remains from the Raddatz Rockshelter, SK5, Wisconsin," *Wisconsin Archeologist* 40, no. 2 (1959): 87–88; Charles Edward Cleland, "Re-analysis of Faunal Remains from Raddatz Rockshelter, Sauk County, Wisconsin," in *The Prehistoric Animal Ecology and Ethnozoology of the Upper Great Lakes Region* (University of Michigan Museum of Anthropology Anthropological Papers no. 29; Ann Arbor: University of Michigan, 1966): 99; Paul W. Parmalee, "Remains of Rare and Extinct Birds from Illinois Indian Sites," *The Auk* 75, no. 2 (April 1958): 172.
12. Henry Ellsworth Cole, *A Standard History of Sauk County, Wisconsin* (Chicago: Lewis, 1918), 6; William H. Canfield, *Outline Sketches of Sauk County*, vol. 1 (Baraboo, WI: A.N. Kellogg, 1861–1874), 11; *History of Crawford and Richland Counties, Wisconsin* (Springfield, IL: Union, 1884), 1238.

13. Alan Watts, *Nature, Man and Woman* (New York: Knopf Doubleday, 2012; first published 1958), 41–42.
14. Alice Kehoe, "Robert Leonard Hall," in *Encyclopedia of Global Archaeology*, edited by Claire Smith (Springer Verlag, 2014).
15. Paul Nabokov, personal communication; Robert L. Hall, *An Archaeology of the Soul: North American Indian Belief and Ritual* (Champaign, IL: University of Illinois Press, 1997); Alexander von Gernet and Peter Timmins, "Pipes and Parakeets: Constructing Meaning in an Early Iroquoian Context," in *Archaeology as Long-Term History*, edited by Ian Hodder (Cambridge: Cambridge University Press), 37–38; Alexander Von Gernet, "New Directions in the Construction of Prehistoric Amerindian Belief System," in *Ancient Images, Ancient Thought: The Archaeology of Ideology*, edited by A. Sean Goldsmith (Proceedings of the Twenty-Third Annual Conference, University of Calgary Archaeological Association, 1992), 133–134.
16. Susan C. Power, *Feathered Serpents and Winged Beings: Early Art of the Southeastern Indians* (Athens: University of Georgia Press, 2004), 22–24; William Snyder Webb and Charles E Snow, *The Adena People* (Lexington: University of Kentucky Reports in Anthropology and Archaeology, 6, 1945): 212–216; William S. Webb and Raymond S. Baby, *The Adena People, no. 2* (Columbus: Ohio Historical Society, 1957), 45.
17. William F. Romain, *Shamans of the Lost World: A Cognitive Approach to the Prehistoric Religion of the Ohio Hopewell* (New York: Rowman & Littlefield, 2009), 7. Ripinsky-Naxon, Michael. *The Nature of Shamanism: Substance and Function of a Religious Metaphor* (Albany, NY: SUNY Press, 1993), 70. LaBarre, Weston. "Shamanic Origins of Religion and Medicine." *Journal of Psychedelic Drugs* vol. 11, no. 1/2 (Jan.–June 1979):9.
18. Michael Harner, "Magic Darts, Bewitching Shamans, and Curing Shamans," in Jeremy Narby and Francis Huxley, *Shamans through Time* (New York: Tarcher/Penguin, 2004), 195; Richard Evans Schultes, Albert Hofmann, and Christian Ratsch, *Plants of the Gods: Their Sacred, Healing, and Hallucinogenic Powers* (Rochester, VT: Healing Arts Press, 2001), 156.
19. Romain, *Shamans*, 20; James A. Brown, "The Archaeology of Ancient Religion in the Eastern Woodlands," *Annual Review of Anthropology* 26 (1997): 468.
20. Peter T. Furst, *Flesh of the Gods: The Ritual Use of Hallucinogens* (New York: Praeger, 1972), ix.
21. Rick Strassman, *DMT: The Spirit Molecule: A Doctor's Revolutionary Research into the Biology of Near-Death and Mystical Experiences* (Rochester, VT: Park Street Press, 2001), chapter 4; Terence McKenna, "Time and Mind" partial transcription of a taped workshop, May 26–27, 1990, New Mexico, online at www.erowid.org/chemicals/dmt/dmt_writings3.shtml, viewed January 19, 2015.
22. Keewaydinoquay. "The Legend of Miskwedo." *Journal of Psychedelic Drugs* vol. 11, no. 1/2 (Jan.–June 1979): 29. La Barre, Weston. "The Pre-Peyote Mescal Bean Cult," in *The Peyote Cult* (5th ed; Norman, Oklahoma: University of Oklahoma Press, 1989), 105-109. La Barre, Weston. "Anthropological Perspectives on Hallucination and Hallucinogens," in *Hallucinations: Behavior, Experience, and Theory* edited

by R.K. Siegel and L.J. West (New York: John Wiley and Sons, 1975), 37. Howard, James H. "Potawatomi Mescalism and Its Relationship to the Diffusion of the Peyote Cult." *Plains Anthropologist* Vol. 7, No. 16 (June 1962), 125 and 133. Richard Evans Schultes, "An Overview of Hallucinogens in the Western Hemisphere," in Peter T. Furst, *Flesh of the Gods: The Ritual Use of Hallucinogens* (New York: Praeger, 1972): 54.

23. Romain, *Shamans*, 20–23; Power, *Feathered Serpents*, 24; Narby and Huxley, *Shamans through Time*, 1–8.
24. Karine Taché, "New Perspectives on Meadowood Trade Items," *American Antiquity* 76, no. 1 (January 2011): 60–61.
25. David W. Penney, "The Adena Engraved Tablets: A Study of Art Prehistory," *Midcontinental Journal of Archeology* 5, no. 1 (1980): 30–35.
26. Webb and Baby, *The Adena People*, 85–92.
27. David W. Penney, "The Archaeology of Aesthetics," in *Hero, Hawk, and Open Hand: American Indian Art of the Ancient Midwest and South*, edited by Richard F. Townsend (Chicago: Art Institute of Chicago; New Haven: Yale University Press, 2004), 43–45; Power, *Feathered Serpents*, 25–28.
28. David M. Strothers and Timothy J. Abel, "Archaeological Reflections on the Late Archaic and Early Woodland Time Periods in the Western Lake Erie Region," *Archaeology of Eastern North America* 21 (Fall 1993): 43.
29. Taché, "New Perspectives," 60; Earl C. Townsend, *Birdstones of the North American Indian: A Study of These Most Interesting Stone Forms, the Area of Their Distribution, Their Cultural Provenience, Possible Uses, and Antiquity* (Indianapolis: Privately printed, 1959), 117; Strothers and Abel, "Archaeological Reflections," 43, 86.
30. Marshall Joseph Becker, "Birdstones: New Inferences Based on Examples from the Area around Waverly, New York," *Bulletin, Journal of the New York State Archaeological Association* 126 (2012): 5–7; Robert Hall, *An Archaeology of the Soul: North American Indian Belief and Ritual* (Urbana: University of Illinois Press, 1997), 115.
31. D. Bruce Dickson, "The Atlatl Assessed: A Review of Recent Anthropological Approaches to Prehistoric North American Weaponry," *Bulletin of the Texas Anthropological Society* 56 (1985): 6, 9, 13; Eric A. Powell, "Paleolithic Pastime," *Archaeology* 62, no. 1 (2009): 16.
32. Dickson, "The Atlatl Assessed," 18–20; Hall, *An Archaeology of the Soul*, 115; Joyce M. Szabo, *Prehistoric Bannerstones and Birdstones of the Midwest with a Catalogue of the Collection from the Ohio Historical Society* (master's thesis, Vanderbilt University, May 1978), 50–53; Townsend, *Birdstones*, devotes an entire chapter to the atlatl grip theory.
33. Robert L. Hall, "An Anthropocentric Perspective for Eastern United States Prehistory," *American Antiquity* 42, no. 4 (October 1977): 504; Townsend, *Birdstones*, 113–117.
34. Alice Beck Kehoe, *America before the European Invasions* (New York: Longman, 2002), 66–68.

35. Richard F. Townsend, "American Landscapes, Seen and Unseen," in *Hero, Hawk, and Open Hand: American Indian Art of the Ancient Midwest and South* (Chicago: Art Institute of Chicago; New Haven: Yale University Press, 2004), 21.
36. Romain, *Shamans*, 42–45; Bradley T. Lepper, "The Newark Earthworks: Monumental Geometry and Astronomy at a Hopewellian Pilgrimage Center," in *Hero, Hawk, and Open Hand*, edited by Richard F. Townsend (Chicago: Art Institute of Chicago; New Haven: Yale University Press, 2004), 77.
37. Power, *Feathered Serpents*, 44–48; Theler and Boszhardt, *Twelve Millennia*, 111.
38. Romain, *Shamans*, 180.
39. Ibid., 199.
40. Thomas E. Emerson, Kenneth B. Farnsworth, Sarah U. Wisseman, and Randall E. Hughes, "The Allure of the Exotic: Reexamining the Use of Local and Distant Pipestone Quarries in Ohio Hopewell Pipe Caches," *American Antiquity* 78, no. 1 (January 2013): 48; Ted J. Brasser, "Self-Directed Pipe Effigies," *Man in the Northeast* 19 (George's Mills, NH: Anthropological Research Center of Northern New England, 1980), 96–97; James A. Brown, "The Shamanic Element in the Hopewell Period Ritual," in *Recreating Hopewell*, edited by Douglas Charles and Jane Buikstra (Gainesville: University Press of Florida, 2006), 481.
41. Johannes Wilbert, *Tobacco and Shamanism in South America* (New Haven: Yale University Press, 1987), 142; Romain, *Shamans*, 179–183; Alexander von Gernet, "Hallucinogens and the Origins of the Iroquoian Pipe/Tobacco/Smoking Complex," in *Proceedings of the 1989 Smoking Pipes Conference: Selected Papers* (Rochester, NY: Rochester Museum and Science Center Research Records no. 22, 1992): 177.
42. Gonzalo Fernandez de Oviedo, quoted in Wilbert, *Tobacco and Shamanism*, 10.
43. Richard Evans Schultes, "An Overview of Hallucinogens in the Western Hemisphere," in Peter T. Furst, *Flesh of the Gods: The Ritual Use of Hallucinogens* (New York: Praeger, 1972): 54; Johannnes Wilbert, "Tobacco and Shamanic Ecstasy among the Warao Indians of Venezuela," in Furst, *Flesh of the Gods*, 56; Louis Nicolas, *The Codex Canadensis and the Writings of Louis Nicolas* (Tulsa, OK: Gilcrease Museum and Montréal: McGill-Queen's University Press, 2011), 274; Joseph-Francois Lafitau, *Customs of the American Indians Compared with the Customs of Primitive Times*, vol. 2 (Toronto: Champlain Society, 1974–1977), 83.
44. Romain, *Shamans*, 179–183; Schultes, Hofmann, and Ratsch, *Plants of the Gods*, 51–55, 110; Richard Evans Schultes and Albert Hofmann, *The Botany and Chemistry of Hallucinogens*, 2nd ed. (Springfield, IL: Charles C Thomas, 1980), 56, 65; Brown, "The Shamanic Element in the Hopewell Period Ritual,"487; Wilbert, "Tobacco and Shamanic Ecstasy among the Warao Indians of Venezuela," 56–57.
45. N'omi B. Greber, "Enclosures and Communities in Ohio Hopewell," in Charles and Buikstra, eds., *Recreating Hopewell*, 77; Dobkin de Rios quoted by Brown in Charles and Buikstra, eds., *Recreating Hopewell*, 483.
46. Robert A. Birmingham and Leslie E. Eisenberg, *Indian Mounds of Wisconsin* (Madison: University of Wisconsin Press, 2000), 101–108; Theler and Boszhardt, *Twelve Millennia*, 127, 138–139.

47. Birmingham and Eisenberg, *Indian Mounds*, 109–115; Theler and Boszhardt, *Twelve Millennia*, 139.
48. Birmingham and Eisenberg, *Indian Mounds*, 116–119.
49. Ibid., *Indian Mounds*, 129.
50. Timothy Pauketat, *Cahokia: Ancient America's Great City on the Mississippi* (New York: Viking, 2009), 10. Other archaeologists disagree with some of Pauketat's claims; I have generally followed his interpretations of the evidence.

Chapter 3

51. Antoine Le Page du Pratz, *The History of Louisiana: Or of the Western Parts of Virginia and Carolina* . . . (London: Becket, 1774), 333, Pierre François Xavier de Charlevoix, *Journal of a Voyage to North-America . . . In a series of letters to the Duchess of Lesdiguieres* . . . (London: R. Goadly, 1763), 312.
52. Alice Beck Kehoe, *America before the European Invasions* (New York: Longman, 2002), 173, 177.
53. Pauketat, *Cahokia*, 2–8.
54. Lucretia S. Kelly, "Patterns of Faunal Exploitation at Cahokia," in *Cahokia: Domination and Ideology in the Mississippian World*, edited by Timothy Pauketat and Thomas Emerson (Lincoln: University of Nebraska Press, 2000), 79, 84–85.
55. Bruce D. Smith, "Middle Mississippi Exploitation of Animal Populations: A Predictive Model," *American Antiquity* 39, no. 2 (April 1974): 280–282, 288–289.
56. J. Hill Hamon, "Bird Remains from a Sioux Indian Midden," *Plains Anthropologist* 6, no. 13 (1961): 208–212.
57. Timothy R. Pauketat and Thomas E. Emerson, "The Ideology of Authority and the Power of the Pot," *American Anthropologist*, n.s., 93, no. 4 (December 1991): 924.
58. William Strachey, *The Historie of Travell into Virginia Britania (1612)* (London: The Hakluyt Society, 1953): 72; Susan Power, *Feathered Serpents & Winged Beings: Early Art of the Southeastern Indians* (Athens: University of Georgia Press, 2004), 70.
59. Power, *Feathered Serpents*, 119–122.
60. Sanchia Schweinsberger, "Bone Flutes and Whistles Found in Ohio Valley Sites," *Proceedings of the Indiana Academy of Sciences* 54 (1949): 28–33.
61. Power, *Feathered Serpents*, 70; Paul W. Parmalee, "Additional Noteworthy Records of Birds from Archaeological Sites," *The Wilson Bulletin* 79, no. 2 (June 1967): 159; Paul W. Parmalee, "Remains of Rare and Extinct Birds from Illinois Indian Sites," *The Auk* 75, no. 2 (April 1958): 172.
62. Richard L. Fishel, "Medicine Birds and Mill Creek-Middle Mississippian Interaction: The Contents of Feature 8 at the Phipps Site (13CK21)," *American Antiquity* 62, no. 3 (July 1997): 550.
63. Pauketat, *Cahokia*, 59–63; John Broihahn, personal communication, August 23, 2014.
64. Power, *Feathered Serpents*, 172.
65. Mark Wagner, Mary R. McCorvie, and Charles Swedlund, "Mississipian Cosmology and Rock Art at the Millstone Bluff Site in Southern Illinois," in *The Rock-Art*

of Eastern North America, edited by Carol Diaz-Granados and James R. Duncan (Tuscaloosa: University of Alabama Press, 2004), 45–50, 62–63.

66. F. Kent Reilly, "People of Earth, People of Sky: Visualizing the Sacred in Native American Art of the Mississippian Period," in *Hero, Hawk, and Open Hand: American Indian Art of the Ancient Midwest and South*, edited by Richard F. Townsend (Chicago: Art Institute of Chicago; New Haven: Yale University Press, 2004), 127–128; Pauketat and Emerson, "The Ideology of Authority," 924–926.
67. Power, *Feathered Serpents*, 169–172; F. Kent Reilly, "The Petaloid Motif: A Celestial Symbolic Locative in the Shell Art of Spiro," in *Ancient Objects and Sacred Realms: Interpretations of Mississippian Iconography*, edited by F. Kent Reilly and James Garber (Austin: University of Texas Press, 2007), 40–42.
68. Pauketat and Emerson, "The Ideology of Authority," 922–924, 929.
69. Power, *Feathered Serpents*, 92–95.
70. James A. Brown, "Spiro Art and Its Mortuary Contexts," in *Death and the Afterlife in Pre-Columbian America*, edited by Elizabeth Benson (Washington, DC: Dumbarton Oaks, 1985), 22.
71. Power, *Feathered Serpents*, 139–140.
72. James A. Brown, "The Cahokian Expression," in *Hero, Hawk, and Open Hand: American Indian Art of the Ancient Midwest and South*, edited by Richard F. Townsend (Chicago: Art Institute of Chicago; New Haven: Yale University Press, 2004), 118–120.
73. Thomas E. Emerson, "Materializing Cahokia Shamans," *Southeastern Archaeology* 22, no. (Fall 2003): 138–140; Kehoe, *America before the European Invasions*, 171.
74. Carol Diaz-Granados, "Marking Stone, Land, Body, and Spirit: Rock Art in Mississippian Iconography," in *Hero, Hawk, and Open Hand: American Indian Art of the Ancient Midwest and South*, edited by Richard F. Townsend (Chicago: Art Institute of Chicago; New Haven: Yale University Press, 2004), , 141–142.
75. Diaz-Granados, "Marking Stone," 142.
76. Charles H. Faulkner, Bill Deane, and Howard H. Earnest Jr. "A Mississippian Period Ritual Cave in Tennessee," *American Antiquity* 49, no. 2 (April 1984): 354, 359; George M. Crothers, "Early Woodland Ritual Use of Caves in Eastern North America," *American Antiquity* 77, no. 3 (July 2012): 532–535.
77. Carol Diaz-Granados and James Richard Duncan, *The Petroglyphs and Pictographs of Missouri* (Tuscaloosa: University of Alabama Press, 2000), 232; Robert Salzer and Grace Rajnovich, *The Gottschall Rockshelter: An Archaeological Mystery* (St. Paul, MN: Prairie Smoke Press, 2000), 3.
78. Pauketat, *Cahokia*, 90–91; James A. Brown, "On the Identity of the Birdman of the Southeastern Ceremonial Complex," in *Ancient Objects and Sacred Realms: Interpretations of Mississippian Iconography*, edited by F. K. Reilly and J. Garber (Austin: University of Texas Press, 2007): 58–59; Salzer and Rajnovich, *The Gottschall Rockshelter*, 31.
79. Paul Radin, *Winnebago Hero Cycles: A Study in Aboriginal Literature* (Baltimore: Waverly Press, 1948), 116–133; Salzer and Rajnovich, *The Gottschall Rockshelter*, 31.

80. Robert Boszhardt and Geri Straub, *Hidden Thunder: Rock Art of the Upper Midwest* (Madison: Wisconsin Historical Society Press, 2016), 148.
81. Pauketat, "The Forgotten History of the Mississippians," in *North American Archaeology*, edited by T. R. Pauketat and D. D. Loren (Oxford: Blackwell, 2005), 191–193.
82. Carol Diaz-Granados and James R. Duncan, *The Petroglyphs and Pictographs of Missouri* (Tuscaloosa: University of Alabama Press, 2013), 234.
83. Brown, "On the Identity of the Birdman," 91.
84. Pauketat, *Cahokia*, 37–38; Pauketat, "The Forgotten History of the Mississippians," 196–200, 207–208.
85. Theler and Boszhardt, *Twelve Millennia*, 157–161; David Overstreet, "Oneota Prehistory and History," *Wisconsin Archeologist* 78, nos. 1–2 (January–December 1997): 251–255; D. W. Benn, "Hawks, Serpents, and Bird-Men: Emergence of the Oneota Mode of Production," *Plains Anthropologist* 34, no. 125 (1989): 233–235.
86. Theler and Boszhardt, *Twelve Millennia*, 201–204; Jodie A. O'Gorman, "Assessing Oneota Diet And Health: A Community and Lifeway Perspective," *Midcontinental Journal of Archaeology* 30, no. 1 (Spring 2005): 119; Amanda Jones, "Faunal Analysis: Reconstructing Subsistency and Seasonality at the Tremaine Site (47LC95)," *University of Wisconsin–La Crosse Journal of Undergraduate Research* 17 (2014): 11–12; Katherine Phyllis Stevenson, "Oneota Subsistence-Related Behavior in the Driftless Area: A Study of the Valley View Site Near La Crosse, Wisconsin" (PhD diss., University of Wisconsin, 1985), 145–148.
87. Douglas H. Ubelaker and Waldo R. Wedel, "Bird Bones, Burials, and Bundles in Plains Archaeology," *American Antiquity* 40, no. 4 (October 1975): 446–449; Paul W. Parmalee, "Additional Noteworthy Records of Birds from Archaeological Sites," *The Wilson Bulletin* 79, no. 2 (June 1967): 157, 160; Warren L. Wittry, "A Raven Headdress from Sauk County, Wisconsin," *Wisconsin Archeologist* 43, no. 4 (1963): 90–93.
88. Benn, "Hawks, Serpents, and Bird-Men," 245–249.
89. Thomas E. Berres, *Power and Gender in Oneota Culture* (Dekalb, IL: Northern Illinois University Press, 2001), 142, 154–162.
90. Benn, "Hawks, Serpents, and Bird-Men," 245–249.
91. Diaz-Granados and Duncan, *The Petroglyphs and Pictographs of Missouri*, 6–7; Minnesota Historical Society, "Jeffers Petroglyphs," Web site at http://sites.mnhs.org/historic-sites/jeffers-petroglyphs/the-rock, visited September 22, 2014.
92. Berres, *Power and Gender*, 149; Robert Salzer, "Wisconsin Rock Art," *Wisconsin Archeologist* 78, nos. 1–2 (1997): 50–53.
93. Diaz-Granados and Duncan, *The Petroglyphs and Pictographs of Missouri*, 28–29, 61–62, 153, 195.
94. James B. Griffin, *The Fort Ancient Aspect* (Ann Arbor: University of Michigan Press, 1943), 46, 73–84, 166, 170, 375; Gwynn A., Henderson, ed., *Fort Ancient Cultural Dynamics in the Middle Ohio Valley* (Madison, WI: Prehistory Press, 1992), 341; Penelope B. Drooker, *The View from Madisonville: Protohistoric Western Fort An-*

cient Interaction Patterns (Ann Arbor: Memoirs of the Museum of Anthropology, University of Michigan, no. 31, 1997), 331.

Chapter 4

95. Mike Williams, *Prehistoric Belief: Shamans, Trance, and the Afterlife* (Stroud, Gloucestershire: The History Press, 2010), 12.
96. Evan T. Pritchard, *No Word for Time: The Way of the Algonquin People* (San Francisco: Council Oak Books, 2001), 61; E. Thomas Mails, *Mystic Warriors of the Plains: The Culture, Arts, Crafts, and Religion of the Plains Indians* (first published 1972; edition quoted, New York: Mallard Press, 1991), 151.
97. Pierre-Charles de Liette, "Memoir of DeGannes Concerning the Illinois Country," *Collections of the Illinois State Historical Library* 23 (1934): 375–380.
98. Paul Radin, *The Winnebago Tribe* (Washington, DC: 37th Annual Report of the Bureau of American Ethnology, 1923), 167.
99. Mails, *Mystic Warriors of the Plains*, 129–132.
100. John Fire, *Lame Deer, Seeker of Visions* (New York: Simon and Schuster, 1994), 6.
101. Radin, *The Winnebago Tribe*, 298.
102. M. Inez Hilger, "Chippewa Child Life and Its Cultural Background," *Bulletin 146, Smithsonian Institution Bureau of American Ethnology* (Washington, DC: GPO, 1951), 44.
103. George Catlin, *Letters and Notes on the Manners, Customs, and Condition of the North American Indians*, vol. 1 (London: The author, 1841), 36–37.
104. Mails, *Mystic Warriors of the Plains*, 134–136; Paul W. Parmalee, "The Avifauna from Prehistoric Arikara Sites in South Dakota," *Plains Anthropologist* 22, no. 77 (August 1977): 214–216.
105. Elizabeth Tooker, *Native North American Spirituality of the Eastern Woodlands* (New York: Paulist Press, 1979), 144–163; Thomas E. Berres, *Power and Gender in Oneota Culture* (Dekalb, IL: Northern Illinois University Press, 2001), 153–154; Paul Radin, *Winnebago Hero Cycles: A Study in Aboriginal Literature* (Baltimore: Waverly Press, 1948), 136; Hilger, "Chippewa Child Life," 45.
106. Mails, *Mystic Warriors of the Plains*, 142, 145; Luther Standing Bear, *My Indian Boyhood* (Lincoln: University of Nebraska Press, 1931), 89.
107. John Heckewelder. *A Narrative of the Mission of the United Brethren . . .* (Cleveland: Burrows Bros. Co., 1907), 178.
108. Standing Bear, *My Indian Boyhood*, 72–73.
109. Jonathan Carver, *Travels through the Interior Parts of North-America in the Years 1766, 1767, and 1768* (London: The author, 1778), 226–227.
110. Pierre-Antoine Tabeau, *Narrative of Loisel's Expedition to the Upper Missouri* (Norman: University of Oklahoma Press, 1939), 90; Standing Bear, *My Indian Boyhood*, 78.
111. Pritchard, *Bird Medicine* (Barnes & Noble ebook edition, 2015), 41–64.
112. Tabeau, *Narrative of Loisel's Expedition*, 90–91.

113. Parmalee, "The Avifauna from Prehistoric Arikara sites . . . ," 207; Standing Bear, *My Indian Boyhood*, 78–83.
114. Standing Bear, *My Indian Boyhood*, 85–86.
115. Giacomo Constantin Beltrami, *A Pilgrimage in America* . . . (London, 1828; edition quoted, Chicago, Quadrangle Books, 1962), 244–245.
116. Catlin, *Letters and Notes*, 164; Rev. Peter Jones, (Kahkewaquonaby), *History of the Ojebway Indians* (London: A.W. Bennett, 1861), 160.
117. Luther Standing Bear, *Land of the Spotted Eagle* (Boston: Houghton Mifflin, 1931), 188; Colin Taylor, *Wapah' a: The Plains Feathered Head-Dress* (Wyk auf Foehr, Germany: Verlag für Amerikanistik, 1996), 25–31.
118. Catlin, *Letters and Notes*, 147–148.
119. Ibid., 146.
120. Josephine Paterek, *Encyclopedia of American Indian Costume* (Denver: ABC-CLIO, 1994), 85, 125, 138; James Otto Lewis, *The Aboriginal Port-folio: A Collection of Portraits of the Most Celebrated Chiefs of the North American Indians* (Philadelphia: J.O. Lewis, 1835–1836); online at www.wisconsinhistory.org/turningpoints/search.asp?id=119; Thomas McKenney and James Hall, *The History of the Indian Tribes of North America. . . Embellished with One Hundred and Twenty Portraits. . .* (Philadelphia: E. C. Biddle, 1836–1844); online at https://content.lib.washington.edu/mckenneyhallweb/; Maximilian zu Wied-Neuwied, *Maximilian Prince of Wied's Travels in the Interior of North America, during the Years 1832–1834* [illustrated by Karl Bodmer] (London: Ackermann, 1843–1844); online at www.gallery.oldbookart.com/main.php?g2_itemId=30172.
121. Reuben G. Thwaites, ed., *The Jesuit Relations and Allied Documents* (Cleveland: Burrows Bros., 1896–1901), vol. 47, 145; Standing Bear, *Land of the Spotted Eagle*, 65; Standing Bear, *My Indian Boyhood*, 88; Henry Schoolcraft, *Narrative Journals of Travels from Detroit Northwest through the Great Chain of American Lakes to the Sources of the Mississippi River in the Year 1820* (Albany, NY: 1821; edition quoted, New York: Arno, 1970), 247–250.
122. John Heckewelder, *History, Manners, and Customs of the Indian Nations Who Once Inhabited Pennsylvania and the Neighboring States* (published 1818; edition quoted, Philadelphia: Historical Society of Pennsylvania, 1876), 202–203; Paterek, *Encyclopedia of American Indian Costume*, 56.
123. George Henry Loskiel, *History of the Mission of the United Brethren among the Indians in North America* (London, 1794), 92; John Long, *Voyages and Travels in the Years 1768–1788* (Chicago: Lakeside Press, 1922), 49; Thwaites, *Jesuit Relations*, 60: 199; Paterek, *Encyclopedia of American Indian Costume*, passim (fans); Catlin, *Letters and Notes*, 242; Standing Bear, *My Indian Boyhood*, 172; Hilger, "Chippewa Child Life," 154, 163.
124. James Kenney, "Journal of . . . [1761–1763]," *Pennsylvania Magazine of History and Biography* 37 (1913): 31–32.
125. Thwaites, *Jesuit Relations*, 59: 127–131, 149; Louis Hennepin, *A New Discovery of a Vast Country in America* (first published in Paris, 1697; edition quoted, Chicago:

A. C. McClurg & Co., 1903), 119, 124–125; Frederick Hodge, *Handbook of American Indians North of Mexico*, vol. 1 (Washington, DC: GPO, 1912), 193.

126. Catlin, *Letters and Notes*, 235.
127. Chretien Le Clercq, *New Relation of Gaspesia with the Customs and Religion of the Gaspesian Indians* (first published in Paris, 1691; edition quoted, Toronto: Champlain Society, 1910), 281.
128. Radin, *The Winnebago Tribe*, 65; Adriaen van der Donck, *A Description of the New Netherlands* (Syracuse, NY: Syracuse University Press, 1968), 51.
129. Loskiel, *History of the Mission*, 92–93.
130. Ibid., 93; Schoolcraft, *Narrative Journals*, 381; Zebulon Pike, *Journals of . . . with Letters and Related Documents*, vol. 1 (Norman, OK: University of Oklahoma Press, 1966), 130.
131. Louis-Armand de Lom d'Arce, baron de Lahontan, *New Voyages to North-America* (London, 1703), 109.
132. John D. Hunter, *Manners and Customs of Several Indian Tribes Located West of the Mississippi . . .* (Minneapolis: Ross and Haines, 1957), 290–291.
133. Chief [Simon] Pokagon, *O-Gi-Maiv-Kiue Mit-I- Gwa-Ki (Queen of the Woods) . . .* (Hartford, MI, 1899), 110–111.
134. Emma Helen Blair, *The Indian Tribes of the Upper Mississippi Valley and Region of the Great Lakes: As Described by Nicolas Perrot . . .*, vol. 1 (Cleveland: Arthur H. Clark Company, 1911), 304–305.
135. Bacqueville de la Potherie, *Histoire de l'Amerique Septentrionale* (Paris, 1722), 81, quoted in *Wisconsin Historical Collections* 16, no. 9 (1902).
136. Claude Dablon, "Relation of What Occurred Most Remarkable in the Missions of the Fathers of the Society of Jesus, . . . in the Years 1671 and 1672," in Thwaites, *Jesuit Relations*, 56: 121.
137. John G. Shea, *Early Voyages Up and Down the Mississippi, by Cavelier, St. Cosme, Le Sueur, Gravier, and Guignas* (Albany, NY, 1861), 103; Pierre Esprit Radisson, *Voyages of . . . , Being an Account of His Travels and Experiences among the North American Indians, from 1652 to 1684* (New York: P. Smith, 1943), 225.
138. Long, *Voyages and Travels*, 70; David Zeisberger, *History of the Northern American Indians* (Columbus, OH, 1910), 119; Pokagon, *O-Gi-Maiv-Kiue Mit-I- Gwa-Ki*, 168–169.
139. Donald Worcester and Thomas F. Schilz, "The Spread of Firearms among the Indians on the Anglo-French Frontiers," *American Indian Quarterly* 8, no. 2 (Spring 1984): 108–111; Roger Carpenter, "Making War More Lethal: Iroquois vs. Huron in the Great Lakes Region, 1609 to 1650," *Michigan Historical Review* 27, no. 2 (Fall 2001).
140. Long, *Voyages and Travels*, 48–49.
141. "An Old Chippewa Chief," *Milwaukee Sentinel*, August 16, 1895.
142. John James Audubon, "Audubon's Journey Up the Mississippi," *Illinois State Historical Society Journal* 35 (1942): 154–155.
143. Thomas Morris, *Miscellanies in Prose and Verse* (London: Printed for James Ridgway, 1791), 15.

144. William Biggs, *Narrative of . . . While He Was a Prisoner with the Kickapoo Indians* (New York: Garland Publishing, 1977), 8–9.
145. Pokagon, *O-Gi-Maiv-Kiue Mit-I- Gwa-Ki*, 130–131; Thomas W. Alford, *Civilization* (Norman: University of Oklahoma Press, 1936): 39–40.
146. Mails, *Mystic Warriors of the Plains*, 96.
147. Luther Standing Bear, *My People, the Sioux* (Boston: Houghton Mifflin, 1928), 39, and *My Indian Boyhood*, 70.
148. Carver, *Travels through the Interior*, 468–469; De Witt Clinton, *Letters on the Natural History and Internal Resources of the State of New York, by Hibernicus* (New York: Bliss and White, 1822), 72.
149. William N. Fenton and Merle H. Deardorff, "The Last Passenger Pigeon Hunts of the Cornplanter Senecas," *Journal of the Washington Academy of Sciences* 33, no. 10 (October 1943): 292.
150. Ibid., 312.
151. Evan T. Pritchard, *Bird Medicine: The Sacred Power of Bird Shamanism* (Rochester, VT: Bear & Co., 2013), 15–16.
152. Ibid., 18.
153. Ibid., 43.

Chapter 5

154. "Journal of the First Voyage of Columbus," in *The Northmen, Columbus and Cabot, 985–1503*, edited by Julius E. Olson and Edward Gaylord Bourne (New York: Charles Scribner's Sons, 1906), 106–107.
155. Edward G. Bourne, *Narratives of the Career of Hernando de Soto in the Conquest of Florida as Told by a Knight of Elvas and in a Relation by Luys Hernandez de Beidma . . .*, vol. 2 (New York: Barnes and Co., 1904), 120–150; George P. Winship, *The Journey of Coronado, 1540–1542, from the City of Mexico to the Grand Canon of the Colorado and the Buffalo Plains of Texas, Kansas and Nebraska, as Told by Himself and His Followers* (New York: Barnes & Co., 1904), x–xi; Consul W. Butterfield, *History of Brulé's Discoveries and Explorations, 1610–1626* (Cleveland: Helman-Taylor Co., 1898), 167–174; Robert Cavelier de La Salle, *Relation of the Discoveries and Voyages of Cavelier de La Salle from 1679 to 1681: The Official Narrative* (Chicago: Caxton Club, 1901), 253–259.
156. Bourne, *Narratives*, 81, 113–114, 120; Winship, *The Journey of Coronado*, 43; the eagle on the prow was probably burnished copper rather than gold, but the Spanish heard what they wanted to hear.
157. Gonzalo Fernandez de Oviedo, "Historia General y Natural de las Indias" (Toledo, 1526), quoted in David B. Quinn, ed., *New American World* (New York, 1979), 1: 263; Jacques Cartier, *The Voyages of . . .* (Ottawa, 1924) 144.
158. Gabriel Sagard, *The Long Journey to the Country of the Hurons by Father Gabriel Sagard* (Toronto: Champlain Society, 1939), xxxiv.
159. Ibid., 33.
160. Ibid., 215.

161. Gabriel Sagard. *Histoire du Canada et Voyages que les Frères Mineurs Recollects y Ont Faicts pour la Conversion des Infidèles depuis l'an 1615* (Paris: Libr. Tross, 1866), 674. Puritan minister Cotton Mather in Boston also couldn't watch birds "without being rapt into admiration of ye Divine Architect" and feel "Admonitions of Piety, and Exhortations to Believe and Adore an Infinite God" ("Unpublished Manuscripts [1712 & 1716] . . . on the Passenger Pigeon," *The Auk* 55 [1938]: 475–476).
162. Sagard, *Long Journey*, 217.
163. Sagard, *Histoire du Canada*, 666.
164. Sagard, *Long Journey*, 190.
165. Sagard, *Long Journey*, 220–221 and 98; *Histoire du Canada*, 666. See previous chapter for the uses of eagle feathers.
166. Sagard, *Long Journey*, 220.
167. Sagard, *Histoire du Canada*, 672–673
168. Ibid., 666.
169. Ibid., 672–673.
170. Ibid., 669–671.
171. Sagard, *Long Journey*, 260
172. Ibid., 244, 270–272.
173. Arthur Edward Jones, "Catalogue of Jesuit Missionaries to New France and Louisiana, 1611 to 1800," in *The Jesuit Relations and Allied Documents*, edited by Reuben G. Thwaites (Cleveland: Burrows Bros., 1896–1901), 71: 120–182.
174. David Hechenberger, "The Jesuits: History and Impact: From Their Origins Prior to the Baroque Crisis to Their Role in the Illinois Country," *Journal of the Illinois State Historical Society* 100, no. 2 (Summer 2007): 88–96; John Walthall and Thomas Emerson, "Indians and French in the Midcontinent," in *Calumet and Fleur-de-Lys: Archaeology of Indian and French Contact in the Midcontinent* (Washington, DC: Smithsonian Institution Press, 1992), 9.
175. Thwaites, *Jesuit Relations*, 12: 15 and 11: 251.
176. Dominique Deslandres, "Exemplo Aeque ut Verbo: The French Jesuits Missionary World," *The Jesuits: Cultures, Sciences, and the Arts, 1540-1773*, edited by John W. O'Malley (Toronto: University of Toronto Press, 1999), 261.
177. Francois-Marc Gagnon, ed. *The Codex Canadensis and the Writings of Louis Nicolas* (Tulsa, OK: Gilcrease Museum and Montreal: McGill-Queens University Press, 2011), 354–355.
178. Ibid., 10–12; Germaine Warkentin, "Aristotle in New France: Louis Nicolas and the Making of the Codex Canadensis," *French Colonial History* 11 (2010): 77.
179. Private communication from archaeologist Robert Birmingham; Claude Allouez, "Journal of Father Claude Allouez's Voyage into the Outaouac Country," in Thwaites, *Jesuit Relations*, 50: 271, 295; Guy Tremblay, "Louis Nicolas, Sa Vie et Son Oeuvre; les Divers Modes de Transport des Indiens Americains" (unpublished master's thesis, University of Montreal, July 1983), 10.
180. Gagnon, *Codex Canadensis*, 13, 19; *Wisconsin Historical Collections* 13: 404–405; Tremblay, "Louis Nicolas," 13.

181. Gagnon, *Codex Canadensis*, 19–20; Warkentin, "Aristotle in New France," 81.
182. Warkentin, "Aristotle in New France," 81–82; Gagnon, *Codex Canadensis*, 18–26; Tremblay, "Louis Nicolas," 24–32.
183. All three were published more than four centuries later, the dictionary as *L'Algonquin au XVIIe siècle: une édition critique, analysée et commentée de la Grammaire algonquine du père Louis Nicolas*, edited by Diane Daviault (Quebec: Presses de l'Universite du Quebec, 1994), and the latter two in Gagnon, *Codex Canadensis*.
184. The earlier works were Francisco Hernández, *Plantas y Animales de la Nueva Espana, y sus virtudes por Francisco Hernandez, y de Latin en Romance por Fr. Francisco Ximenez* (Mexico, 1615; Latin editions published 1628, 1648); *Histoire Naturelle des Indes: The Drake Manuscript in the Pierpont Morgan Library* (New York: W.W. Norton, 1996); and John Josselyn, *New-England's Rarities Discovered in Birds, Beasts, Fishes, Serpents, and Plants of That Country . . .* (London: G. Widdowes, 1672; reprinted Boston: William Veazie, 1865).
185. Gagnon, *Codex Canadensis*, 265. This summary is the subtitle Nicolas gave to his manuscript.
186. Ibid., 266.
187. Louis Nicolas, *Codex Canadensis*, Plate 52, Late 1600s - early 1700s, Ink on paper, Overall: 13 1/4 × 8 1/2 in. (33.7 × 21.6 cm), GM 4726.7.052, Gilcrease Museum, Tulsa, Oklahoma.
188. Ibid., 303, 348–349.
189. Ibid., 303.
190. Ibid., 303.
191. Ibid., 349.
192. Ibid., 349.
193. Ibid., 354–355.
194. Gagnon, *Codex Canadensis*, 354–355.
195. John Gyles, *Memoirs of Odd Adventures, Strange Deliverances, etc . . .* (Boston: Kneeland and Greene, 1736), 36.
196. Nicolas, *Codex Canadensis*, 356.
197. Ibid., 367–368.
198. Ibid., 369–370.
199. Gagnon, *Codex Canadensis*, 364.
200. Gagnon, *Codex Canadensis*, 363–366. There were only four Joliets in Canada at the time, the famous explorer Louis Joliet and his three brothers, as listed in *Dictionnaire Généalogique des Familles du Québec* by René Jetté (Montréal: Presses de l'Université de Montréal, 1983), 603.
201. Claude Allouez, "Of the Mission St. Francois Xavier and the Nations Dependent Thereon," in Thwaites, *Jesuit Relations*, 55: 195–197; Nicolas, *Codex Canadensis*, 370
202. Conrad Gesner, *Icones Avium Omnium, quae in Historia Avium Conradi Gesneri Describuntur . . .*(Zurich, 1555); Francois Du Creux, *Historiae Canadensis, sev Novae-Franciae Libri Decem . . .* (Paris, 1664); Gagnon, *Codex Canadensis*, 15–17, 63–65.

203. Tremblay, "Louis Nicolas," 37; Warkentin, "Aristotle in New France," 90.
204. Gagnon, *Codex Canadensis*, 3–4, 28; Warkentin, "Aristotle in New France," 90, 99.
205. Warkentin, "Aristotle in New France," 99; Gagnon, *Codex Canadensis*, 3.
206. Tremblay, "Louis Nicolas," 25.
207. Louis Vivier, "Letter from Father Vivier, Missionary among the Ilinois, to Father * * * ," in Thwaites, *Jesuit Relations*, 69: 141.
208. Thwaites, *Jesuit Relations*.
209. Allouez, "Of the Mission," 215.
210. Jacques Marquette, "Of the Mission of Saint Esprit at Chagaouamigong Point," in Thwaites, *Jesuit Relations*, 54: 187.
211. Sebastien Rasles, "Letter from Father Sebastien Rasles, Missionary of the Society of Jesus in New France, to Monsieur his Brother," in Thwaites, *Jesuit Relations*, 67: 166–167.
212. Vivier, "Letter from Father Vivier," 143.
213. Jacques Marquette, "Of the First Voyage Made by Father Marquette toward New Mexico, and How the Idea Thereof Was Conceived," in Thwaites, *Jesuit Relations*, 59: 87.
214. Ibid., 87.
215. Ibid., 107.
216. Ibid., 107; Gagnon, *Codex Canadensis*, plate 55.
217. This map, never published, is in the archives of the Depot des Cartes et Plans de la Marine in Paris. Shea, *Early Voyages*, 66–67. Lucien Campeau's article, "Les Cartes Relatives à la Découverte du Mississipi par le P. Jacques Marquette et Louis Jolliet," in *Les Cahiers des Dix*, no. 47 (1992): 41–92 (online at http://wihist.org/2woqJMk), reproduces Franquelin's map and Joliet's sketch of the monster.
218. Natalia Maree Belting, "The Piasa: It Isn't a Bird!" *Journal of the Illinois State Historical Society* (1908–1984) 66, no. 3 (Autumn 1973): 302–305.
219. Marquette, "Of the First Voyage," 147–149.
220. Louis Joliet, "Letter 10 October 1674," photostat of original manuscript in the Wisconsin Historical Society Archives (SC 283; facsimile at www.wisconsinhistory.org/turningpoints/search.asp?id=14).
221. Louis Hennepin, *Father Louis Hennepin's Description of Louisiana, Newly Discovered to the Southwest of New France* . . . (Minneapolis: Minnesota Historical Society, 1938): 45–46.

Chapter 6

222. Allen G. Debus, *Man and Nature in the Renaissance* (Cambridge, UK: Cambridge University Press, 1978), 12–13.
223. George McKinnon Wrong and William Stewart Wallace, eds., *Review of Historical Publications Relating to Canada*, vol. 1 (Toronto: Univ. of Toronto Press, 1897), 39–40; Thwaites, *Jesuit Relations*, 59: 292–298; Melchisedec Thevenot, "Découverte de quelques pays et nations de l´Amerique Septentrionale [par le P. Marquette]," in *Recueil de Voyages de Mr Thevenot* . . . (Paris: Estienne Michallet, 1681).

224. Celine Dupre, "Cavelier de La Salle, Rene-Robert," *Dictionary of Canadian Biography Online* (www.biographi.ca/, viewed December 25, 2012).
225. Milo M. Quaife, ed., *The Western Country in the Seventeenth Century: The Memoirs of Lamothe Cadillac and Pierre Liette* (Chicago: Lakeside Press, 1947), 130.
226. Antoine de Lamothe Cadillac, "Description de la Riviere du Detroit," in Pierre Margry, *Decouvertes et Etablissements des Francais dans l'Ouest et dans le Sud de l'Amerique Septentrionale (1614–1754): Memoires et Documents Originaux*, vol. 5 (Paris: 1871–1886), 193–194.
227. *Relation of the Discoveries and Voyages of Cavelier de La Salle*, 137–139.
228. Their harrowing odyssey is recounted in Henri Joutel, *Journal Historique du Dernier Voyage que Feu M. de LaSale Fit dans le Golfe de Mexique* (Paris: Chez E. Robinot, 1713; English translation the following year).
229. David M. Hayne, "Lom D'Arce de Lahontan, Louis-Armand de, Baron de Lahontan," *Dictionary of Canadian Biography Online* (www.biographi.ca/, viewed Dec. 30, 2012).
230. Louis-Armand de Lom d'Arce, baron de Lahontan, "Memoirs of North America," in *New Voyages to North America*, vol. 1 (Chicago: McClurg & Co., 1905), 299–300.
231. Ibid., 355–356.
232. Ibid., 354.
233. Ibid., 109.
234. Ibid., 353–355.
235. James Smith, *An Account of the Remarkable Occurrences in the Life and Travels of . . . during his Captivity with the Indians . . .* (Cincinnati, 1870): 57–58. This and other cases of supposed transmutation of species are discussed in Frank G. Speck and John Witthoft, "Some Notable Life-Histories in Zoological Folklore," *Journal of American Folklore* 60, no. 238 (1947): 347.
236. Smith, *An Account of the Remarkable Occurrences*, 83–84.
237. Jonathan Carver, *Journals of . . . and Related Documents, 1766 1770* (Minneapolis: Minnesota Historical Society Press, 1976), 15–21.
238. Jonathan Carver, *Travels through the Interior Parts of North-America in the Years 1766, 1767, and 1768* (London: Walter and Crowder, 1778), 441.
239. Ibid., 526.
240. Ibid., 208–210, 215.
241. Ibid., 467.
242. Ibid., 466–469.
243. Ibid., vii.
244. Wayne Bodle, "Review of *Gilbert Imlay: Citizen of the World* by Wil Verhoven" in *William and Mary Quarterly* (April 2009): 445–449.
245. Gilbert Imlay, *A Topographical Description of the Western Territory of North America* (London: Debrett, 1792), 94–95.
246. Ibid., 220–228. Imlay was referring to Mark Catesby's *The Natural History of Carolina, Florida, and the Bahama Islands . . .* (London, 1731); because Catesby traveled only on the East Coast, his work is not discussed here.

247. Georges-Louis Leclerc, Comte de Buffon, *Histoire Naturelle, Générale et Particulière, avec la Description du Cabinet du Roi* (Paris: L'Imprimerie Royale: 1749–1788); Keith Thomson, *Jefferson's Shadow: The Story of His Science* (New Haven, CT: Yale University Press, 2012), 280; Kevin Hayes, *The Road to Monticello: The Life and Mind of Thomas Jefferson* (New York: Oxford University Press, 2008), 201; Elsa Allen, "The History of American Ornithology before Audubon," *Transactions of the American Philosophical Society*, n.s., 41, part 3 (1951): 532.
248. John F. Kennedy, "Remarks at a Dinner Honoring Nobel Prize Winners of the Western Hemisphere," April 29, 1962. Gerhard Peters and John T. Woolley, *The American Presidency Project*, www.presidency.ucsb.edu/ws/?pid=8623.
249. Jefferson to George Rogers Clark, December 4, 1783, in the Draper Manuscripts (George Rogers Clark Papers, 52J, 93–95) at the Wisconsin Historical Society (www.americanjourneys.org/aj-140/).
250. David Lavender, "Appendix I: Jefferson's Instructions to Lewis," in *The Way to the Western Sea* (Lincoln: University of Nebraska Press, 2001), 389–394; Silvio Bedini, "The Scientific Instruments of the Lewis and Clark Expedition," *Great Plains Quarterly* 4, no. 1 (1984): 54–69; Lavender, *The Way to the Western Sea*, 66.
251. Paul A. Johnsgard, *Lewis and Clark on the Great Plains: A Natural History* (Lincoln: University of Nebraska Press, 2003), 5.
252. *Journals of the Lewis and Clark Expedition*, edited by Gary Moulton (Lincoln: University of Nebraska Press, 1993), 4: 450–451.
253. Ibid., 4: 460–462. Curly brackets in the original.
254. Ibid., 3: 82–84; Johnsgard, *Lewis and Clark on the Great Plains*, 52.
255. *Journals of the Lewis and Clark Expedition*, 3: 179.
256. Johnsgard, *Lewis and Clark on the Great Plains*, 3–5.
257. Quoted in Carla Mulford, *The Cambridge Companion to Benjamin Franklin* (New York: Cambridge University Press, 2009), 119–120.
258. Mulford, *Cambridge Companion to Benjamin Franklin*, 120.
259. John Heckewelder, *A Narrative of the Mission of the United Brethren . . .* (Cleveland: Burrows Bros., 1907), 76–77.

Chapter 7

260. John James Audubon, "Account of the Method of Drawing Birds Employed by J. J. Audubon, Esq., F.R.S.E. in a Letter to a Friend," *Edinburgh Journal of Science* 8, no. 1 (Jan. 1828): 48.
261. John James Audubon, *Writings and Drawings* (New York: Library of America, 1999), 794.
262. Francis H. Herrick, *Audubon the Naturalist: A History of His Life and Time*, (New York: Appleton, 1917), 1: xxxix–liv; Richard Rhodes, *John James Audubon: The Making of an American* (New York: Knopf, 2004), 3–5; Audubon, *Writings and Drawings*, 861–869.
263. Henry David Thoreau, "Natural History of Massachusetts," in *Excursions* (Boston: Ticknor and Fields, 1863), 37.

264. Audubon, *Writings and Drawings*, 46–48.
265. John James Audubon, *Ornithological Biography, or An Account of the Habits of the Birds of the United States of America . . .*, vol. 1 (Edinburgh: Adam Black, 1831–1839), 466. The role of Audubon's editors is discussed in Christoph Irmscher's "Audubon the Writer," a talk delivered at the July 25, 2007, premiere of the PBS American Masters documentary *John James Audubon: Drawn from Nature*; viewed online October 17, 2014, at www.audubonroyaloctavos.com/SITE/pages/Irmscher .html.
266. Audubon, "Account of the Method," 54.
267. Andrew J. Lewis, *A Democracy of Facts: Natural History in the Early Republic* (Philadelphia: University of Pennsylvania Press, 2011), 7; "Second Charta: 22 April 1663," in Henry G. Lyons, *The Royal Society, 1660-1940* (Cambridge, UK: The University Press, 1944), 329.
268. Leonard Warren, *Maclure of New Harmony: Scientist, Progressive Educator, Radical Philanthropist* (Bloomington: Indiana University Press, 2009), 20; Charlotte M. Porter, *The Eagle's Nest: Natural History and American Ideas, 1812-1842* (Tuscaloosa, AL: University of Alabama, Press, 1986), 2–5; George Ord, "A Memoir of Thomas Say," in *The Complete Writings of Thomas Say on the Entomology of North America*, edited by John L. Le Conte (New York: Balliere Brothers, 1859), ix.
269. Roger Hahn. *The Anatomy of a Scientific Institution: the Paris Academy of Sciences, 1666-1803*. (Berkeley: University of California Press, 1971), 36; Lewis, *A Democracy of Facts*, 11–12; Patricia T. Stroud, *Emperor of Nature: Charles-Lucien Bonaparte and His World* (Philadelphia: University of Pennsylvania Press, 2000), 53.
270. Herrick, *Audubon: The Naturalist*, 1: 225, 328, 362–363, and 2: 87.
271. Audubon, *Ornithological Biography*, 1: 438; Janet Haven, "Alexander Wilson" on the University of Virginia's Web site, *Alexander Wilson, American Ornithologist*, at http://xroads.virginia.edu/~public/wilson/bio.html, viewed October 19, 2014.
272. Laura Rigal, "Empire of Birds: Alexander Wilson's American Ornithology," *Huntington Library Quarterly* 59, nos. 2/3 (1996): 238–239.
273. Biographical details from: Haven, "Alexander Wilson"; Michael Ziser, "Alexander Wilson," in *Early American Nature Writers: A Biographical Encyclopedia* (Westport, CT: Greenwood, 2008), 392; Elliott Coues, "Bibliographical Appendix: List of Faunal Publications Relating to North American Ornithology," in *Birds of the Colorado Valley* (Washington, DC: Department of the Interior, US Geological Survey of the Territories, Miscellaneous Publications no. 11; Government Printing Office, 1878), 600.
274. Audubon, *Ornithological Biography*, 1: 438–439.
275. Herrick, *Audubon the Naturalist*, 1: 224–225.
276. Gregory H. Nobles, "Ornithology and Enterprise: Making and Marketing John James Audubon's 'The Birds of America,'" *Proceedings of the American Antiquarian Society* 113, pt. 2 (2003): 275–278, 287–294.
277. Herrick, *Audubon the Naturalist*, 1: 224–228, 329; 2: 87.
278. Lewis, *A Democracy of Facts*, 3–7.

279. John T. Battalio, *The Rhetoric of Science in the Evolution of American Ornithological Discourse* (Stamford, CT: Ablex, 1998), 29; Coues, "Bibliographical Appendix," 601–658.
280. Raymond P. Stearns, *Science in the British Colonies of America* (Urbana: University of Illinois Press, 1970), 6; Battalio, *The Rhetoric of Science*, 26–29; Tod Highsmith, email to the author, September 7, 2016.
281. Audubon, "Account of the Method of Drawing Birds," 50, 53.
282. John Moring, *Early American Naturalists: Exploring the American West, 1804-1900* (New York: Cooper Square Press, 2002), 74–76; Ord, "Memoir of Thomas Say," vii–ix.
283. Ord, "Memoir of Thomas Say," xiii–xx; Benjamin Coates, *A Biographical Sketch of the Late Thomas Say, Esq.: Read Before the Academy of Natural Sciences of Philadelphia, December 16, 1834* (Philadelphia: W.P. Gibbons, 1835), 22–24.
284. Biographical details from Ord, "Memoir of Thomas Say," xviii; Porter, *The Eagle's Nest*, 100–108; Coates, *A Biographical Sketch*, 3; Stroud, *Emperor of Nature*, 43–48.
285. Porter, *The Eagle's Nest*, 91–93; William Stanton, *American Scientific Exploration, 1803–1860*, American Philosophical Society Web site at http://www.amphilsoc.org/guides/stanton/0335.htm, viewed Oct. 26, 2014.
286. Stroud, *Emperor of Nature*, 48–49; Stanton, *American Scientific Exploration*.
287. Charles-Lucien Bonaparte, *American Ornithology, or the Natural History of Birds Inhabiting the United States, Not Given by Wilson* (Philadelphia: Carey & Lea, 1825), 1: 7; Ord, "Memoir of Thomas Say," xviii; Porter, *The Eagle's Nest*, 100–108; Coates, *A Biographical Sketch*, 3; Stroud, *Emperor of Nature*, 62.
288. Coates, *A Biographical Sketch*, 23.
289. Charles Lucian Bonaparte, *The Genera of North American Birds, and a Synopsis of the Species Found within the Territory of the United States . . .* (New York: J. Seymour, 1828), 8; Charles Lucian Bonaparte, "Observations on the Nomenclature of Wilson's Ornithology," in *Journal of the Academy of Natural Sciences of Philadelphia* 4, no. 1 (1824): 168; Stroud, *Emperor of Nature*, 63.
290. Stroud, *Emperor of Nature*, 78, 91–94, 128–129.
291. Coues, *Bibliographical Appendix*, 610–617.
292. Elias Durand, *Memoir of the Late Thomas Nuttall* (Philadelphia: Sherman and Son, 1860), 4–5; Henry Marie Brackenridge, *Journal of a Voyage up the River Missouri in 1811*, edited by Reuben G. Thwaites, in *Early Western Travels, 1748–1846*, vol. 6 (originally published Baltimore, 1816; Cleveland: Burrows Brothers, 1904) 102; Richard Henry Dana, *Two Years Before the Mast* (New York: Harper Brothers, 1842), 360.
293. Durand, *Memoir of the Late Thomas Nuttall*, 5–7.
294. Thomas Nuttall, "Travels into the Old Northwest: An Unpublished 1810 Diary," *Chronica Botanica* 14 (1950–51): 81–82; Thomas Nuttall, "Remarks and Inquiries Concerning the Birds of Massachusetts," *Memoirs of the American Academy of Arts and Sciences*, n.s., 1 (Cambridge: Charles Folsom, 1833): 105–106.
295. Thomas Nuttall, *Manual of the Ornithology of the United States and of Canada*, 2 vols. (Cambridge: Hilliard and Brown, 1832 and 1834), 1: v.

296. Nuttall, *Manual*, 1: 19
297. Jeannette E. Graustein, *Thomas Nuttall, Naturalist: Explorations in America, 1808–1841* (Cambridge, MA: Harvard University Press, 1967), 379–396; American Ornithologists' Union, *AOU Checklist of North and Middle American Birds*, viewed online October 25, 2014, at http://checklist.aou.org/.
298. Donald M. Hassler, "Henry Rowe Schoolcraft," in *Early American Nature Writers: A Biographical Encyclopedia* (Westport, CT: Greenwood, 2008), 311–315; Milo M. Quaife, *Chicago and the Old Northwest, 1673–1835* (Chicago: University of Chicago Press, 1913), 453.
299. William Cooper, "Description of a New Species of Grosbeak, Inhabiting the Northwestern Territory of the United States," in *Annals of the Lyceum of Natural History of New York*, vol. 1, pt. 2 (1825): 219–222; Michael J. Mossman, "H. R. Schoolcraft and Natural History on the Western Frontier, Part 4: Indian Agency Years with Thomas McKenney," *Passenger Pigeon* 55 (1993): 156.
300. Jeremiah Joyce, *A Familiar Introduction to the Arts and Sciences for the Use of Schools and Young Persons . . .* (London: Longman Hurst, 1819), 299.
301. Benjamin Smith Barton, *Fragments of the Natural History of Pennsylvania* (Philadelphia: Way & Groff, 1799), vi; Elliott Coues, *Birds of the Colorado Valley* (Washington, DC: GPO, 1878), 374; Lewis, *A Democracy of Facts*, 31–34.
302. Dana, *Two Years before the Mast*, 361.
303. Elliott Coues, *Field Ornithology. Comprising a Manual of Instruction for Procuring, Preparing and Preserving Birds, and a Check List of North American Birds* (Salem [MA]: Naturalists' Agency, 1874), 5.
304. Paul Russell Cutright and Michael J. Brodhead, *Elliott Coues: Naturalist and Frontier Historian* (Urbana: University of Illinois Press, 2001), 269–271; Joel Asaph Allen, *Biographical Memoir of Elliott Coues, 1842–1899* (Washington, DC: National Academy of Sciences, 1909): 402–404.
305. Coues, *Field Ornithology*, 4.
306. Ibid., 8–9.
307. Ibid., 34.
308. Ibid., 36.
309. Ibid., 53–54, 58.
310. Ibid., 27.
311. Ibid., 30.
312. Edward Lee Greene, *Pittonia: A Series of Papers Relating to Botany and Botanists* (Berkeley, CA: 1887–1889), 253; Publius Lawson, "Thure Kumlien," *Transactions of the Wisconsin Academy of Sciences, Arts, and Letters* 20 (Madison: The Academy, 1921): 667.
313. Lawson, "Thure Kumlien," 669; Angie Kumlien Main, "Studies in Ornithology at Lake Koshkonong . . . ," *Transactions of the Wisconsin Academy of Sciences, Arts and Letters* 37 (1945): 100–101.
314. Angie Kumlien Main, "Thure Kumlien, Koshkonong Naturalist. Part III," *Wisconsin Magazine of History* 27, no. 3 (March 1944): 323; Lawson, "Thure Kumlien,"

663, 672–673, 678; Mrs. H. J. Taylor, "Thure Ludwig Theodor Kumlien," *The Wilson Bulletin* 48, no. 2 (June 1936): 88.

315. Lawson, "Thure Kumlien," 669; Main, "Thure Kumlien, Koshkonong Naturalist," 335; Taylor, "Thure Ludwig Theodor Kumlien," 90.
316. Taylor, "Thure Ludwig Theodor Kumlien," 90; Lawson, "Thure Kumlien," 683; Main, "Thure Kumlien, Koshkonong Naturalist," 340–341.
317. Lawson, "Thure Kumlien," 683–684; Main, "Thure Kumlien," 342.
318. Battalio, *The Rhetoric of Science*, 29; American Ornithologists' Union, *Fifty Years' Progress of American Ornithology, 1883–1933*, rev. ed. (Lancaster, PA: AOU, 1933), 131.
319. American Ornithologists' Union, *Fifty Years' Progress*, 29; J. A. Allen, *Biographical Memoir, Elliott Coues 1842-1899, Read Before the National Academy of Sciences, April 1909* (Washington, DC: National Academy of Sciences, 1909), 401–409.
320. Quoted in Mark V. Barrow, *A Passion for Birds: American Ornithology After Audubon* (Princeton, NJ: Princeton University Press, 2000), 89, 94.
321. "Reply to the Preceding Communication; Containing Observations on Certain Species of Game Birds, and on the Names by Which They Are Distinguished among Sportsmen," *Medical Repository* 8, no. 2 (1804): 124–128; Gurdon Trumbull, *Names and Portraits of Birds which Interest Gunners* (New York: Harper, 1888), 80–83.
322. Edwin Muir, "The Animals," in *The Complete Poems of Edwin Muir* (Aberdeen, UK: Association for Scottish Literary Studies, 1991), 196.

Chapter 8

323. Charles A. Eastman, *Indian Boyhood* (New York: Doubleday, Page & Co., 1915), 5, 53.
324. Ibid., 49, 155.
325. Brent Berlin, *Ethnobiological Classification* (Princeton, NJ: Princeton University Press, 1992), 31–34.
326. Cecil H. Brown, *Language and Living Things: Uniformities in Folk Classification and Naming* (New Brunswick, NJ: Rutgers University Press, 1984); Brent Berlin and John P. O'Neill, "The Pervasiveness of Onomatopoeia in Aguaruna and Huambisa Bird Names," *Journal of Ethnobiology* 1, no. 2 (December 1981): 238–261.
327. Brent Berlin, "The First Congress of Ethnozoological Nomenclature," *Journal of the Royal Anthropological Institute* 12 (2006): 523–544.
328. Claude E. Schaeffer, "Bird Nomenclature and Principles of Avian Taxonomy of the Blackfeet Indians," *Journal of the Washington Academy of Sciences* 40 (1950): 37–46; Laurence Irving, "Stability in Eskimo Naming of Birds on Cumberland Sound, Baffin Island," *University of Alaska Anthropological Papers* 10 (1961/63): 1–12.
329. Marius Barbeau, "The Language of Canada in the Voyages of Jacques Cartier (1534–1538)," *National Museum of Canada Bulletin No. 173* (Ottawa, 1961): 108–229.
330. Gabriel Theodat Sagard, *Dictionaire de la Langue Huron* (Paris, 1632), reprinted in vol. 4 of his *Histoire du Canada et Voyages . . .* (Paris: Librarie Tross, 1866); Sagard, *Long Journey to the Country of the Hurons* (Ottawa: Champlain Society, 1939), 219;

John L. Steckley, *Words of the Huron* (Waterloo, ON: Wilfrid Laurier University Press, 2007), xv; Victor Egon Hanzeli, *Missionary Linguistics in New France* (The Hague: Mouton, 1969); Claude Allouez, "Recit d'un 3e Voyage Faict aux Ilinois," in *The Jesuit Relations and Allied Documents*, edited by Reuben G. Thwaites (Cleveland: Burrows Bros., 1896–1901), 60: 161.

331. Louis-Armand, baron de Lahontan, "A Short Dictionary of the Most Universal Language of the Savages," *New Voyages to North America* (London, 1703), edited by Reuben Gold Thwaites (Chicago: McClurg, 1905), 2: 732–750.
332. Lists of 17th-century Algonquian bird names are recorded in Quinn (appendix B), Gilliam, 1–2, Hariot, 139–150, Swanton, 96–99, Wood, 117–124, and Williams, 113–118.
333. Thomas Hariot, *A Briefe and True Report of the New Found Land of Virginia . . .* (London, 1588; reprinted in Quinn), 3: 139-155; John R. Swanton, "Newly Discovered Powhatan Bird Names," *Journal of the Washington Academy of Sciences* 24 (1934): 96–99.
334. Roger Williams, *A Key into the Language of America* (London, 1643), reprinted in vol. 1 of Publications of the Narragansett Club (1866), 113–118.
335. The following discussion is based on A. L. Pickens, "A Comparison of Cherokee and Pioneer Bird-Nomenclature," *Southern Folklore Quarterly* 7 (1943): 213–221; and Arlene Fradkin, *Cherokee Folk Zoology: The Animal World of a Native American People, 1700–1838* (New York: Garland, 1990).
336. Based on analysis of the current AOU checklist of North American birds.
337. Hanzeli, *Missionary Linguistics*, 100.
338. Hamilton Tyler, *Pueblo Birds and Myths* (Norman: University of Oklahoma Press, 1979), xiii–xiv.
339. Tyler, *Pueblo Birds and Myths*.
340. Sonia Tidemann and Andrew Gosler, *Ethno-Ornithology: Birds, Indigenous Peoples, Culture and Society* (Washington, DC: Earthscan, 2010), 34; Henry Thoreau, *Journal*, vol. 1 (Boston: Houghton Mifflin, 1906), 253.
341. Jacques Cartier, *The Voyages of . . . , Published from the Originals . . . .*, edited by H. P. Biggar (Ottawa, 1924), 144.
342. Louis Hennepin, *A New Discovery of a Vast Country in America* (Chicago: A.C. McClurg, 1903), 109.
343. Capt. John Knox, *An Historical Journal of the Campaigns in North America for the Years 1758 1760* (Toronto: 1914), 1: 317.
344. Nicolas Claude de Fabri, "Notes on Specimens and Pictures of Specimens Brought by the Sieur de Monts from Acadia [1605–1606]," in *New American World*, edited by David B. Quinn (New York: Arno Press and Hector Bye, 1979), 4: 357.
345. Henri Joutel, *Journal of La Salle's Last Voyage* (New York: Corinth Books, 1962), 72.
346. Jean Pierre Aulneau, "[Letters from Canada, 1734-36]," in Arthur Jones, *Rare or Unpublished Documents II: The Aulneau Collection, 1734–1745* (Montreal, 1893), 10; Antoine Simon Le Page du Pratz, *The History of Louisiana, or of the Western Parts of Virginia and Carolina . . .* (London: 1774; edition quoted, New Orleans, 1947), 269.

347. John Francis McDermott, *A Glossary of Mississippi Valley French, 1673–1850* (St. Louis, MO: Washington University Studies, New Series, No. 12, December 1941), 13, 42, 108.
348. Mark Catesby, *The Natural History of Carolina, Florida, and the Bahama Islands* . . . (London, 1731; edition quoted, Chapel Hill: Univ. of North Carolina Press, 1985), 102; Samuel Hearne, *Journey from Prince of Wales's Fort in Hudson's Bay to the Northern Ocean, 1769, 1770, 1771, 1772* (London, 1795; edition quoted, Toronto: Macmillan, 1958), 268; Peter Kalm, *Travels Into North America* (London, 1772; edition quoted, Barre, MA: The Imprint Society, 1972), 245; Sir Charles Blagden, "Letters . . . to Sir Joseph Banks on American Natural History and Politics, 1776–1780," *Bulletin of the New York Public Library* 7, no. 11 (1903): 411.
349. William J. Rupp, "Bird Names and Bird Lore Among the Pennsylvania Germans," in *Proceedings and Addresses, Pennsylvania-German Society* 52, no. 2 (1947): 26, and passim. The conclusions that follow are based on a statistical analysis of bird names found in Rupp's 337-page monograph.
350. Gurdon Trumbull, *Names and Portraits of Birds Which Interest Gunners, with Descriptions in Language Understanded of the People* (New York: Harper Brothers, 1888), vi–vii.
351. Waldo Lee McAtee, *Local Names of Migratory Game Birds*, US Department of Agriculture Miscellaneous Circular no. 13 (Washington, DC: Government Printing Office, 1923).
352. Ibid., 21, 56, 64.
353. Ibid., 19, 23, 64.
354. Ibid., 7, 49.
355. E. R. Kalmbach, "In Memoriam: W.L. McAtee," *The Auk* 80 (1963): 485; W. L. McAtee, *Nomina Abitera* (privately printed, 1945), 1; McAtee, *Nomina Abitera, Supplement* (privately printed, 1954), 7.
356. McAtee, *Nomina Abitera*, 26.
357. Ibid., 40, and McAtee, *Nomina Abitera, Supplement*, [11].
358. McAtee, *Nomina Abitera*, 35.
359. McAtee's manuscript and card files, safely microfilmed, are now in the Department of Manuscripts and University Archives at Cornell University.
360. Richard Severo, "Roger Peterson, 87, the Nation's Guide to the Birds, Is Dead," *New York Times*, July 30, 1996.
361. McAtee, *Local Names*, 32.

Chapter 9

362. Daniel L. Thomas and Lucy B. Thomas, *Kentucky Superstitions* (Princeton, NJ: Princeton University Press, 1920), items 3511, 3513, and 3638–3639; William J. Rupp, "Bird Names and Bird Lore among the Pennsylvania Germans," *Proceedings and Addresses of the Pennsylvania German Society* 52 (1946): 252.
363. The paragraphs that follow are based largely on Fanny D. Bergen, "Animal and Plant Lore," *Memoirs of the American Folk-Lore Society* 7 (1899); Joseph Medard

Carriere, *Tales from the French Folklore of Missouri* (Chicago: Northwestern University Press, 1937); H.H.C. Dunwoody, "Weather Proverbs," *Signal Service Notes*, no. 9 (Washington, DC: War Department, 1883); Edward B. Garriott, "Weather Folk-Lore and Local Weather Signs," *U.S. Dept. of Agriculture Bulletin* (1903); Newbell Niles Puckett, *Folk Beliefs of the Southern Negro* (Chapel Hill: University of North Carolina Press, 1926); Rupp, "Bird Names and Bird Lore," and Thomas and Thomas, *Kentucky Superstitions*.

364. Wolfgang Mieder, Stewart A. Kingsbury, and Kelsie B. Harder, *A Dictionary of American Proverbs* (New York: Oxford University Press, 1992): 10–12.
365. Bartlett Jere Whiting, *Early American Proverbs and Proverbial Phrases* (Cambridge, MA: Harvard University Press, 1977), 32–33.
366. Joel Chandler Harris, *Uncle Remus: His Songs and His Sayings* (New York: Appleton, 1881), 149–152.
367. US Census Bureau, "Table 1. Urban and Rural Population: 1900 to 1990" (Washington, DC: US Census Bureau, October 1995); viewed May 3, 2015, at www.census.gov/population/censusdata/urpop0090.txt.
368. Dunwoody, "Weather Proverbs," 35–36; Garriott, "Weather Folk-Lore," 18.
369. Rupp, "Bird Names and Bird Lore," 247; Dunwoody, "Weather Proverbs," 35; Bergen, "Animal and Plant Lore," item 463.
370. Dunwoody, "Weather Proverbs," 40–41; Rupp, "Bird Names and Bird Lore," 243, 246.
371. Bergen, "Animal and Plant Lore," items 774–775, 778; Thomas and Thomas, *Kentucky Superstitions*, item 3611.
372. Rupp, "Bird Names and Bird Lore," 254; Puckett, *Folk Beliefs*, 362.
373. Rupp, "Bird Names and Bird Lore," 262; Puckett, *Folk Beliefs*, 267–268, 314, 459, 491; Thomas and Thomas, *Kentucky Superstitions*, items 493, 496, 3532.
374. Rupp, "Bird Names and Bird Lore," 262; Thomas and Thomas, *Kentucky Superstitions*, items 130, 139, 3532.
375. Puckett, *Folk Beliefs*, 314, 354, 459, 491.
376. Bergen, "Animal and Plant Lore," 333, 338; Thomas and Thomas, *Kentucky Superstitions*, items 3519–3520, 3653; Rupp, "Bird Names and Bird Lore," 256; Puckett, *Folk Beliefs*, 487–488.
377. Dunwoody, "Weather Proverbs," 38; Puckett, *Folk Beliefs*, 482–483, 488; Thomas and Thomas, *Kentucky Superstitions*, item 3627.
378. Bergen, "Animal and Plant Lore," items 286 and 288; Thomas and Thomas, *Kentucky Superstitions*, item 3607; Rupp, "Bird Names and Bird Lore," 258.
379. *1830, 1880 and 1900 Census of the United States*, via the Historical Census Browser at the University of Virginia (http://mapserver.lib.virginia.edu/); Puckett, *Folk Beliefs*, 7, 31–35.
380. Puckett, *Folk Beliefs*, 18, 168, 469, 489; Richard M. Dorson, *American Negro Folktales* (New York: Fawcett, 1967), 14.
381. Puckett, *Folk Beliefs*, 356; Melville Herskovits, *The Myth of the Negro Past* (New York: Harper Brothers, 1941), 235–260.

382. Robert Tallant, *Voodoo in New Orleans* (New York: Macmillan, 1946), 247.
383. Puckett, *Folk Beliefs*, 120, 150, 156.
384. Ibid., 195, 206.
385. Ibid., 227, 232.
386. Tallant, *Voodoo in New Orleans*, 184.
387. Puckett, *Folk Beliefs*, 290, 297.
388. Tallant, *Voodoo in New Orleans*, 176.
389. Roger Mitchell, "Farm Talk from Marathon County," in *Wisconsin Folklore*, edited by James P. Leary (Madison: University of Wisconsin Press, 1998), 91.
390. Puckett, *Folk Beliefs*, 208; Tallant, *Voodoo in New Orleans*, 184.
391. Hayden White, *The Content of the Form* (Baltimore: Johns Hopkins Press, 1987), 24; Jerome Bruner, *Acts of Meaning* (Cambridge, MA: Harvard University Press, 1990), 72–77, 83.
392. J. Mason Brewer, *Dog Ghosts and Other Texas Negro Folk Tales* (Austin: University of Texas Press, 1958), 50, quoted in Lawrence Levine's *Black Culture and Black Consciousness: Afro-American Folk Thought from Slavery to Freedom* (New York: Oxford University Press, 1977), 113.
393. Robert Boszhardt and Geri Straub, *Hidden Thunder: Rock Art of the Upper Midwest* (Madison: Wisconsin Historical Society Press, 2016), 143.
394. "Why the Buzzard Is Bald," in Tristram P. Coffin, *Indian Tales of North America; an Anthology for the Adult Reader* (Philadelphia: American Folklore Society, 1961), 142–143.
395. Edith Fowke, *Folktales of French Canada* (Toronto: NC PRess Ltd., 1979): 100–101.
396. "Lazy Buzzard," in Richard M. Dorson, *Negro Folktales in Michigan* (Cambridge, MA: Harvard University Press, 1956), 45–46.
397. Martha Strudwick Young, *Plantation Bird Legends* (New York: Appleton, 1902; edition quoted, 1916), 154–156.
398. Ernest W. Baughman, *Type and Motif-Index of the Folktales of England and North America*, Indiana Folklore Series no. 20 (The Hague: Mouton & Co, 1966), xvi–xvii; Richard M. Dorson, *Bloodstoppers & Bearwalkers: Folk Traditions of the Upper Peninsula* (Cambridge, MA: Harvard University Press, 1952), 69.
399. Dorson, *Bloodstoppers & Bearwalkers*, 75.
400. Baughman, *Type and Motif-Index*, xvi–xvii; Richard Dorson, *Jonathan Draws the Long Bow* (Cambridge, MA: Harvard University Press, 1946), 110.
401. Charles E. Brown, *Ben Hooper Tales: Settler's Yarns from Green and Lafayette Counties, Wisconsin* (Madison: Wisconsin Folklore Society, 1944), 5; Baughman, *Type and Motif-Index*, 54.
402. Vance Randolph, *We Always Lie to Strangers: Tall Tales from the Ozarks* (New York: Columbia University Press, 1951), 96–97.
403. Michael Edmonds, *Out of the Northwoods: The Many Lives of Paul Bunyan* (Madison: Wisconsin Historical Society Press, 2009); Charles E. Brown, *Sourdough Sam, Paul Bunyan's Illustrious Chief Cook and Other Famous Culinary Artists of his Great Pinery Logging Camps, Old Time Tales of Kitchen Wizards, the Big Cook Shanty, the*

Camp Fare, the Dinner Horn, and Sam's Cook Book (Madison: Wisconsin Folklore Society, 1945), 2; Charles E. Brown, *Paul Bunyan Natural History* (Madison, WI: Charles E. Brown, 1935), 6.

Chapter 10

404. *The Story of American Hunting and Firearms* (New York: Outdoor Life /Dutton, 1976), 40, 64, 69.
405. Oliver Johnson, "A Home in the Woods; Oliver Johnson's Reminiscences of Early Marion County," *Indiana Historical Society Publications* 16, no. 2 (1951): 194–196.
406. Ibid., 194–196.
407. Ibid., 198–199.
408. Patrick Campbell, *Travels in the Interior Inhabited Parts of North America in the Years 1791 and 1792* (Toronto: Champlain Society, 1937), 173.
409. Isaac Weld Jr., *Travels through the States of North America, and the Provinces of Upper and Lower Canada, in the Years 1795, 1796 and 1797* (London: J. Stockdale, 1799), 335–336.
410. William N. Blane, *An Excursion through the United States and Canada during the Years 1822–23, by an Englishman* (London: Baldwin, Cradock, and Joy, 1824), 173–174.
411. Jonathan Carver, *Travels through the Interior Parts of North-America in the Years 1766, 1767, and 1768* (London: Walter and Crowder, 1778), 470; Timothy Flint, *Recollections of the Last Ten Years, Passed in Occasional Residences and Journeyings in the Valley of the Mississippi* (Boston: Cummings, Hilliard, and Co., 1826), 89.
412. Rose S. Taylor, "Peter Schuster: Dane County Farmer (Part II)," *Wisconsin Magazine of History* 28, no. 4 (June 1945): 450–451.
413. Blane, *An Excursion*, 173–174.
414. H. Clay Merritt, *The Shadow of a Gun* (Chicago: Peterson Co., 1904), 46–47.
415. Sebastien Rasles, "The Wanderings of . . . , 1689–1723," in William I. Kip, *The Early Jesuit Missions in North America* (New York, 1846; Carlisle, Mass.: Applewood Books, 2010), 39; John G. Shea, *Early Voyages Up and Down the Mississippi, by Cavelier, St. Cosme, Le Sueur, Gravier, and Guignas* (Albany, Joel Munsell, 1861), 64.
416. Peter Pond, "1740–75: Journal of . . ." *Wisconsin Historical Collections* 18 (1908): 332–333.
417. Georges-Henri-Victor Collot, *A Journey in North America, Containing a Survey of the Countries Watered by the Mississippi, Ohio, Missouri, and Other Affluing Rivers . . .* (Paris: A. Bertrand, 1826), 1: 166; Fortescue Cuming, *Sketches of a Tour to the Western Country, through the States of Ohio and Kentucky . . . , Early Western Travels, 1748–1846*; vol. 4 (originally published Baltimore, 1816; Cleveland, 1904), 4: 290.
418. Michael Smith, *A Geographical View of the Province of Upper Canada . . .* (Hartford, Conn.: John Russell, 1813), 31–32; William H. Keating, *Narrative of an Expedition to the Source of St. Peter's River, Lake Winnepeek, Lake of the Woods, etc., Performed in the Year 1823 . . .* (Philadelphia: Carey, 1824), 169.

419. Edmund Dana, *Geographical Sketches on the Western Country, Designed for Emigrants and Settlers . . .* (Cincinnati: Looker, Reynolds & Company Printers, 1819), 262; Merritt, *Shadow of a Gun*, 187.
420. Henry William Herbert, *Frank Forester's Field Sports of the United States, and British Provinces, of North America* (New York: Springer & Townsend, 1849), 20.
421. Merritt, *Shadow of a Gun*, 11.
422. Bob Hinman, *The Golden Age of Shotgunning* (New York: Winchester Press, 1971), 92–93.
423. Merritt, *Shadow of a Gun*, 212–213, 227.
424. Jack Musgrave, "Market Hunting in Northern Iowa," *Annals of Iowa* 26, no. 3 (January 1945): 173–175.
425. Ibid., 176.
426. Ibid., 177.
427. Ibid., 180–181.
428. Ibid., 173.
429. Ibid., 183.
430. Ibid., 188–189.
431. US Census Office, *Report on Population of the United States at the Eleventh Census, 1890, Part II* (Washington, DC: GPO, 1895–97), 304–337, 454.
432. Merritt, *Shadow of a Gun*, 41.
433. Ibid., 46–47, 110.
434. Ibid., 69–70.
435. Ibid., 73–75.
436. Ibid., 222–225.
437. Ibid., 221–225.
438. Ibid., 221–225.
439. Herbert, *Frank Forester's Field Sports*, 20.
440. Charles Hallock, *The Sportsman's Gazetteer and General Guide . . .* (New York: Forest and Stream, 1877), 79.
441. Theodore Sherman Palmer, *Hunting Licenses: Their History, Objects, and Limitations*, USDA Biological Survey Bulletin No. 19 (Washington, DC: US Dept. of Agriculture, 1904), 18–19, 55–60.
442. Musgrave, "Market Hunting," 185.
443. Ibid., 180, 185–186.
444. Ibid., 184–185.
445. Mark H. Davis, "Market Hunters vs. Sportsmen on the Prairie," *Minnesota History* (Summer 2006): 50.
446. Kurkpatrick Dorsey, *The Dawn of Conservation Diplomacy: U.S.-Canadian Wildlife Protection Treaties in the Progressive Era* (Seattle: University of Washington Press, 1998), 175–176; "Supplement: The Present Wholesale Destruction of Bird-Life in the United States," *Science*, 7, no. 160 (February 26, 1886): 191–205.
447. "Supplement: The Present Wholesale Destruction," 192–195.

448. Ibid., 194, 196–197.
449. Ibid., 203.
450. Dorsey, *Dawn of Conservation Diplomacy*, 178; William Hornaday, "The Destruction of our Birds and Mammals: A Report on the Results of an Inquiry," Second Annual Report of the New York Zoological Society (New York: The Society, March 15, 1898), 79, 95–96.
451. Hornaday, "The Destruction of our Birds and Mammals," 112–113.
452. American egret, American raven, American scaup duck, Carolina parakeet, curlew, eskimo curlew, golden plover, Hudsonian curlew, Hudsonian godwit, ivory-billed woodpecker, long-billed curlew, northern raven, passenger pigeon, pileated woodpecker, pinnated grouse, ruffed grouse, sandhill crane, snowy egret, trumpeter swan, white pelican, white-necked raven, whooping crane, wild turkey, wood duck, and woodcock.
453. William T. Hornaday, *Our Vanishing Wild Life: Its Extermination and Preservation* (New York: C. Scribner's Sons, 1913), 42–52.
454. John Muir, "Wild Wool," *Overland Monthly* 14, no. 4 (April 1875): 361; Genesis 1:28 (King James Version).
455. John A. Hosteder, "Toward Responsible Growth and Stewardship of Lancaster County's Landscape," in *Pennsylvania Mennonite Heritage* (July 1989): 2–10.
456. Luther Standing Bear, *Land of the Spotted Eagle* (Lincoln: University of Nebraska Press, 2006), 38; quoted in *American Indian Myths and Legends*, edited by Richard Erdoes and Alfonso Ortiz (New York: Knopf Doubleday, 2013), 5.
457. Mark Barrow, *A Passion for Birds: American Ornithology After Audubon* (Princeton, NJ: Princeton University Press, 2000), 112, 118, 129–131; William Dutcher, "Report of the National Association of Audubon Societies . . . [including] a History of the Audubon Movement," *Bird-Lore* 7, no. I (Jan.–Feb. 1905): 56.
458. Theodore Sherman Palmer, *Chronology and Index of the More Important Events in American Game Protection*, USDA Bureau of Biological Survey Bulletin no. 41 (Washington, DC: US Government Printing Office, 1912): 11–12, 17; David Willis, Charles Scalet, and Lester D. Flake, *Introduction to Wildlife and Fisheries* (New York: W.H. Freeman, 2009), 349–350.
459. Neltje Blanchan, *Bird Neighbors: An Introductory Acquaintance with One Hundred And Fifty Birds Commonly Found in the Gardens, Meadows, and Woods about Our Homes* (New York: Doubleday & McClure, 1898); Gloria Shearin, "Neltje Blanchan," in *Early American Nature Writers*, edited by D. Patterson, R. Thompson, and S. Bryson (Westport, CT: Greenwood Press, 2008), 62–69.
460. Searches in OCLC's WorldCat and Readex's *America's Historical Newspapers* databases, Aug. 2, 2015.

Chapter 11

461. Anais Nin, *Seduction of the Minotaur* (Chicago: Swallow Press, 1961), 124.

Sources and Acknowledgments

462. Elliott Coues, "Dr. Coues' Column," *The Osprey, An Illustrated Monthly Magazine of Popular Ornithology* 2 (November 1897): 39-40; Coues, "Bibliographical Appendix: List of Faunal Publications Relating to North American Ornithology," in *Birds of the Colorado Valley* (Washington, DC: Department of the Interior, U.S. Geological Survey of the Territories, Miscellaneous Publications no. 11; Government Printing Office, 1878), 567–1066. Coues's bibliography is still the best guide to the first 350 years of writing about American birds.

Index

Page references in **bold** refer to illustrations.

About the Author

JOEL HEIMAN

Michael Edmonds has been a recreational birder for three decades. He is the author of two award-winning books from the Wisconsin Historical Society Press, *Out of the Northwoods* and *Risking Everything*, and has written several articles for the *Wisconsin Magazine of History* and other journals. He graduated from Harvard University in 1976, joined the Wisconsin Historical Society staff in 1982, and has taught at the University of Wisconsin since 1986.